HOGWILD

A Back-to-the-Land Saga

text and photography by

JOCK LAUTERER

with additional photography by Maggie Lauterer

Other books by the author:

Only in Chapel Hill

Wouldn't Take Nothin' for My Journey Now

Runnin' on Rims

For my children,

Selena, who was raised in a barn;
and Jonathan, who was born because of one.

Appalachian Consortium Press

The Appalachian Consortium is a non-profit educational organization comprised of institutions and agencies located in the Southern Highlands. Our members are volunteers who plan and execute projects which serve 156 mountain counties in seven states. Among our goals are:

- Preserving the cultural heritage of Southern Appalachia
- Protecting the mountain environment
- Publishing manuscripts about the region
- Improving educational opportunities for area students and teachers
- Conducting scientific, social, and economic research
- Promoting a positive image of Appalachia
- Encouraging regional cooperation

The member institutions of the Appalachian Consortium are:

Appalachian State University
Blue Ridge Parkway
East Tennessee State University
Gardner-Webb College
Great Smoky Mountains Natural History Association
John C. Campbell Folk School
Lees-McRae College
Mars Hill College
Mayland Community College
N.C. Division of Archives and History
Southern Appalachian Highlands Conservancy
Southern Highlands Handicraft Guild
U.S. Forest Service
Warren Wilson College
Western Carolina University
Western North Carolina Historical Association

Copyright © 1993 by Appalachian Consortium Press and Jock Lauterer
All rights reserved.

Library of Congress Cataloging-in-Publication Data.
Lauterer, Jock
  HOGWILD: A Back to the Land Saga
  ISBN 0-913239-69-0
  1. Homesteading. 2. Homebuilding-Southern Appalachians. 3. The 1970's—Autobiography. 4. Back-to-the-Land Movement—North Carolina—Autobiography.

    I. Title

HOGWILD

A Back-to-the-Land Saga

CONTENTS:

Foreword ix
A Word to the Wise xvi
Acknowledgments xviii

CHAPTERS:

1 Beginnings / 1
2 How Firm a Foundation / 8
3 Summer of '74: Barn-Razing and Cabin-Raising / 24
4 Roofing: In Which We Become Acquainted with 5,500 Shingles / 36
5 Joisting Along / 45
6 Wherein Our Hero is Floored and Builds a Playhouse / 52
7 In Which We Chink and Daub, and Win a Press War / 72
8 The Cabin Grows as Friends Pitch In / 82
9 In Which We Move a Smokehouse and Create Another Hogwildling / 90
10 Of a Giving House and Lofty Ambitions / 108
11 In Which We Have a Son and are Illuminated / 126
12 Wherein the King of Siam Builds a Chimney / 136
13 Rocking On / 144
14 Hearth and Homing In / 155
15 In Which the Sweatmore Construction Co. Gets Fancy / 166
16 In the Home Stretch / 177
17 Moving In At Last / 190
18 In Which We Live the Dream / 197
19 Life and Death Within the Dream / 203
20 Homecoming / 210

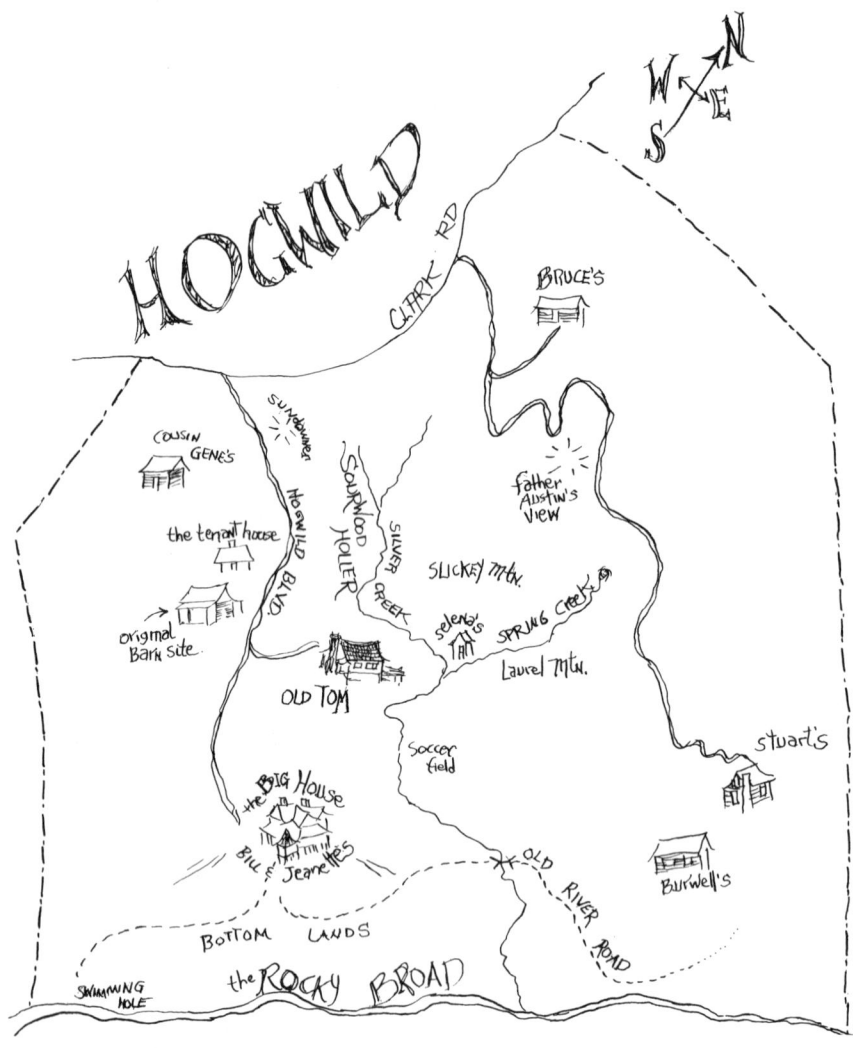

Foreword

*"Talk of mysteries! Think of our life in nature —
daily to be shown matter, to come in contact with it —
rocks, trees, wind on our cheeks! the solid earth!
the actual world! the common sense! Contact! Contact!
Who are we? Where are we?"*

— Henry David Thoreau
Walden

Hogwild. It was an outrageous name for an outrageous notion: that several families might form a loose confederation of homesteaders, buy a big chunk of land, build our own homes and become as self-sufficient as practically possible.

The early 1970s was a time of a quiet green revolution in the backwoods of America. In the front lines were the former campus activists of the '60s, the retired demonstrators for Civil Rights, the tired protesters against the Viet Nam war, and the disillusioned political critics of a federal government perceived as increasingly and inherently adversarial, if not downright evil.

We had repaired to the country where we turned our considerable energy loose on a smaller, more manageable scale. If we couldn't save the world, then we'd work on improving our own world in a small and intensely personal way; we would make our lives harmonious with nature and the physical world as we found it.

It was a notion that fired our times and our minds with a righteous zeal and a sense of profound purpose. We went back to basics. Back to the land.

This is the story of how we joined with five other families on 300 acres in a dream of building our own homes with virtually no previous carpentry experience. It was, as the place came to be known, Hogwild.

Located just east of the rocky flanks of the Blue Ridge Mountains in western North Carolina's Rutherford County, we could have the best of both worlds: the isolation of the mountains yet without the temperature extremes of the high country immediately west of us.

The view west from Hogwild toward Hickory Nut Gorge.

Philosophically and in practice, we were not so much a commune as we were a community; a spontaneous rural assembly of like-minded young coots with families to raise, houses to build, and dreams to try. And we weren't afraid to be a little bit outrageous.

However, we were nothing like the stereotypical version of the '70s commune popularized by Hollywood as inhabited by chronically stoned pot-heads stumbling around obliviously in the woods, mumbling nothing more profound than "cool, man," and sharing everything from property to non-exclusive affections.

At Hogwild, we each owned our own land and lived within the traditional contexts of our marriages. As a community, we had log house-raisings, skinny-dipping in the river, frolicking nights of country music of our own making, weddings in the woods, and parties on the lawn. We shared our tools, our muscles, our time, and the land that was Hogwild.

We were bound together by our common interest in organic foods, our dedication to conservation and environmental sanity, and our desire to provide our children with an appropriate place in which to grow up.

Spiritually and philosophically as builders of our own shelter and caretakers of our own land (as opposed to owners), we were most influenced by Eastern thought and by Native American gentleness. We joked about belonging to the Southern Buddhist Church. But that would have offended my bearded ancestors in the Friends Meeting House, so I concluded somewhat in jest that if I had to reduce my faith to a phrase, then it would be Zen-Quaker. With maybe a touch of Cherokee on the side.

If we were a part of a movement or a cause or a national phenomenon, we certainly didn't know it. There was no newsletter for whoever or whatever we were becoming: "Homesteaders" or so-called "Back-to-the-Landers." We at Hogwild simply felt we were bound together on a private journey in wood.

If we were pioneers, we were unwitting ones; I doubt that pioneers grasp their own uniqueness anyway. That perspective is a luxury of a hindsight we couldn't have at that time. What is clear now is that we were caught up in the creative confluence of a synergy that resulted in some amazing houses and some equally amazing times.

Like others, I found quiet inspiration and sustenance from publications — not so counter-culture as new culture — such as the *Whole Earth Catalog*, and *Mother Earth News*, (which we scoffed at for its "Hey Gang, you can build a house of mudpies in a day" brand of gee-whiz journalism).

Most importantly, Eliot Wigginton's first *Foxfire* book (1972) with its chapter on log cabin building, Eric Sloane's *An Age of Barns* and Art Boericke's *The Woodbutcher's Art, the Guide to Handmade Houses*, all combined to free me from my conventional thinking of how a place should or could be built.

From the very beginning, log cabins have been in my genes. My great-grandfather Nixon Rush, the Quaker minister, was born in the log house (1836) that his father built east of Shepperds Mountain in Randolph County, N.C.

Old Rush was an adventurer who took off for the Colorado Silver Rush of 1859 before settling the family farm, "Rush Hill," in rural Indiana. His son, my grandfather Charles, restored a log cabin in the '20s in Indiana for a summer retreat away from his career as a university librarian.

As a boy, I built "houses" out of orange crates, packing boxes and old tin sheeting, and these propped-up knee-high shanties had the proclivity of collapsing about me with a great clamor as I crawled in.

This is basic instinct, says Bradford Angier, in his fine book, *Building Your Own Cabin in the Woods*. Angier suggests that the impulse of building is so intrinsic to the human spirit that we see it reflected in the role-play of small boys — building clubhouses, tree houses — any sort of slapdash place to call their own.

As a boy growing in the confines of a university town during those formative years, my link to all things woodsy was Scouting in general

Campus folksingers of the '60s, "Jock and Maggie" are ready by 1974 to put songs and dreams into the reality of wood, rock, and mortar.

Bill and Jeanette Byers, the "King and Queen of Hogwild," with daughter, Hannah, in the winter of 1974.

and Troop 39 in particular. The outfit provided me with a confidence in myself and my abilities to live comfortably in a wild setting. Plus, Scouting provided me with the male role modeling and the sort of wilderness experiences that my single parent mother couldn't give me. Troop 39, not so coincidentally, was a pioneering troop, specializing in the skills of lashing pine logs into 20-foot-tall signal towers with an elaborate rope monkey bridge strung from the ground up to one tower, across to the other tower, and down to the ground on the far side.

To an 11-year-old pudgy kid, this edifice took on DeMilleian scale. As an adolescent, I became something of a professional Scout, as was the custom in our troop, becoming a leader of younger boys, wise to the ways of the woods, earning my Eagle rank and staying with the troop until I was 16. Building the cabin was the ultimate expression and extension of my Scouting experience.

But still, none of this would have happened had it not been for Maggie — the girl who bowled me over at a University of North Carolina at Chapel Hill campus meeting where students were taking their turns introducing themselves invariably as Susie Q. from Charlotte or John J. from Raleigh, and so forth — cities so big we didn't even need to be reminded what state they were in... And when it came her turn, this raven-haired beauty stood up and announced simply: "I'm Maggie Palmer, and I'm from the Mountains!"

She was from a town so small back up in the mountains that she knew no one had ever heard of Crossnore (pop. 250) and she wasn't about to be humiliated by all those city slickers.

From the first time I saw her proudly proclaiming her "mountainess," I was, as my friend Tom Callison says, "all eat up."

We became folksingers together, fell in love and were wed. From the start we shared the dream of going to the mountains and having a weekly newspaper.

By 1973 we had realized this dream.

Maggie and I had been married for seven years. Our wildling daughter, Selena, was five. The weekly newspaper, *This Week*, started by two friends and I in 1969, was flourishing. Without knowing it, we had become: YURPies (Young Upwardly-mobile Rural Professionals).

The decision to go Hogwild was the result of myriad influences. One of those events was my cathartic interview for a job as a photographer with *National Geographic* magazine in November 1973. After looking at my work and at several year's worth of bound volumes of *This Week*, the magazine's director of photography, Robert Gilka, told me *This Week* was one of the finest papers of its kind he'd ever seen, and he asked me if I would leave it to move to D.C. Whether the question was rhetorical or specific, it didn't matter.

Leave my paper?! How could I do that and live in Washington, D.C.? Within me, a country mouse vs. city mouse debate raged briefly. But in the end it was no contest. My course as a country editor was already plotted and I suspect Mr. Gilka knew that very well. Yet the funny thing was, I got the distinct impression from the wistful look on Gilka's face as he leafed through my papers that he would have liked to have been in my shoes — and that he wouldn't have left his blooming country weekly in the hills of the Southern Appalachians either.

In that winter of 1973-'74, with our growing awareness of the earth's finite resources and our second house just having received the horrific moniker of a "total electric home," we both began to become susceptible to a drastic alteration in our means of shelter.

We began to feel that we were little more than caretakers and financial trustees of our present, energy-hogging house with its big drafty halls, high ceilings, leaky windows, and a fireplace with such a draw that it could suck all the heat out of the house as fast as the baseboard units could crank it out.

It is said that creative change for the better is stimulated by dissatisfaction with one's present state, and we were ripe for the revolution.

Maggie and I began to suspect that there were basic rules of shelter dynamics we'd never dreamed of before. Here's one:

Lauterer's Law of Site Specifics: That your shelter and the particular dynamics, configuration, location, style, mood, lighting — all combine to form a total ethos that dictates to a great extent the behavior of its inhabitants.

Or, to borrow a turn of phrase from the health folk: you are where you live.

So, we wondered, what if the owner-builder could be active instead of passive, as is usually the case in our dwellings, and dictate his own life and style by fashioning and forming a structure up and around that philosophical and intellectual core long before the first napkin-blueprint is ever drawn.

This was the basic precept of Hogwild, and this, more than cinderblocks, was the foundation for our house. What a thought: that you don't build houses and stick people in them willy-nilly; but rather you take a person or a family and build the house around them as an extension of their personal and corporeal identities. No wonder the house's gestation period was the same as a high school or college undergraduate education. We didn't build this cabin so much as we raised it and nurtured it.

Long before the first shovelful of dirt was moved, before the first nail pounded or the first log laid, this house was sensed — much as a mother-to-be can sense the tiny seed of the embryo long before its presence is officially confirmed. We felt this house deep down, dreamed of it, envisioned it to be the vital, vibrant, nurturing, creation-inspiring and life-supporting homeplace that would itself became a living entity.

Such a house was Tom.

Not that he had a name, much less a gender, early on. But surely the old barn possessed an innate personality: honesty, authenticity, durability and sort of a stout hardiness. So that even as we saw him that first blustery winter day in 1974, all bedraggled with vines, Tom still beckoned.

During the four years it took to build Tom, a lot was going on professionally as well as personally. For one thing, I was losing my hair, and after a couple of years of combing what has been called "a dime's worth of hair over a dollar's worth of head," I decided to take matters in my own hands, so to speak, and I shaved my head. Still do.

At the paper, we became recognized as the best non-daily newspaper in the state. And during this period, we purchased a press and bought out our competition as well.

At the same time, I was beginning to understand the importance of gathering and printing oral history and the development of my brand of what I call "folkloric photojournalism." Those weekly features resulted in two books, *Wouldn't Take Nothin' for My Journey Now* in 1980 (the University of North Carolina Press) and *Runnin' on Rims* in 1986 (Algonquin Books of Chapel Hill).

Without quite realizing it, I came to live the style of those about whom I was writing. Building the cabin gradually evolved into the ultimate expression of who I was, who we were corporally, of our adopted countryness, and of my work.

Professionally these years encompassed a tremendously prolific and creative body of work as a photographer and writer. I can't help suspecting that Hogwild's energy had something to do with that "good karma," as we called it.

As co-publisher and co-editor at *This Week*, I had the luxury of rambling on (and on) in weekly columns in which I regularly proselytized for the homesteading way of life while writing about Tom's progress. Readers came to know Tom as sort of an extension of their own house. "How's the cabin comin'?" they'd say to me on the street, and: "Read about it in the paper." Folks seemed to enjoy it, although fellow journalists kidded me unmercifully for that.

"You're the only reporter I know who can put up one log and get 20 inches of copy out of it," jabbed my buddy Roy Thompson of the *Winston-Salem Journal*. I reckon I deserved it, but every hour I wasn't working on the paper I spent out there in the woods with Old Tom. Were we obsessed? How can I answer that? At lunch we drew designs and "blueprints" on napkins and check receipts. While covering city council meetings I doodled fanciful drawings of cabins amid the notes of subdivision regulations and zoning appeals.

For four years we were immersed in a project that demanded our very best if we were to realize its completion. And we were totally committed to that dream. It took all the extra energy we had — but at the same time, the expenditure sustained and nurtured us.

From the very start, Maggie had the foresight and vision to hint broadly that I should keep a diary of the house building by giving me a journal whose blank pages challenged me as much as the cleared building site. If I could commit mortar, rock, and log to the land, then I could do the same with ink to paper.

What follows are the Chronicles of Tom; the saga of a homesteading couple whose lives were forever changed by those times. Every house we've ever lived in since has borne the indelible stamp of Old Tom and of a place called Hogwild.

A Word to the Wise

"What does not destroy me, makes me stronger."
— Friedrich Nietzsche

To those would-be woodbutchers who turn here for sustenance: building a house from scratch is much like climbing a major mountain — defining "major" as being 14,000 feet or over, a peak so formidable that it will kill you if you go at it blithely unprepared.

First, you have its picture burned in your mind's eye. Early on, when you are deciding to climb it, that is your only vision. Maybe it's a photograph in *National Geographic*, a two-dimensional, four-color separation representing a critical mass, height, wind, cold, deprivation, oxygen starvation, precipice, muscle failure, cold sweat, and fear. At that point, you don't know what you're looking at, and it's a good thing.

You collect your stuff, marshal your forces, make the ten thousand arrangements, set aside the time, and gird your loins. You have an inkling that this will be a once-in-a-lifetime experience, something to tell the grandkids about. The romance of it enthralls you but you really haven't any idea of what you're getting into — except that deep down you know you need to do this thing.

The Tenderfoot sets out. You have your vision to warm you as you draw closer to the mountain and you are gay and glib. After traveling for days, while still 100 miles away, you see its forbidding icy spire against the sky. Impossible. It can't be that big. Nothing can be that big. Something's wrong. We must be there already. Something's wrong with the map, or the odometer, or something.

Your mind gapes at the mountain's monstrosity.

And still you drive toward it, and still it looms, no larger, no nearer. Just a maleficent pyramid jutting violently into a benign baby-blue sky.

You drive through the night — so that the next day when you have arrived at the mountain's base, you can no longer see its top. Instead, inglorious and trivial matters occupy you. Getting to base camp, pitching camp, even the life-sustaining acts of eating and drinking seem somehow remote. A sense of unreality descends, for you can no longer see your goal though you are standing on its very flanks. You are too close to see clearly but too far along to turn back.

As you set out the next morning, you rely on your original vision to sustain you. You climb and climb. Progress seems marginal. Why is this taking so long? No one told me there'd be so many tiny infuriating obstacles: altitude sickness, the skree slope, cutting steps in the snow and ice, the wind, and always the fatigue. In the fine dust of the skree slope you take one step and slide back two. The higher you go and the closer to the triumph, the slower you go. The thin air permits one step at a time. But still, you slog on.

As the air thins, breathing itself becomes a challenge. Even though you may be climbing with buddies, as the labor becomes more intense, you become more insular, bargaining for progress: I must be getting higher if the air is becoming thinner. So where's the freaking end to all this? But no summit greets your pained sight.

Someone has been kind enough to warn you about the "false summit syndrome" in which, because of the mountain's curvature and your position on its side relative to the top, the true summit keeps itself hidden and always seems to recede and indeed elude you — so that you find yourself straining repeatedly for what you think is the Big Top, only to get there and find to your total dismay, another slope stretching 2,000 feet up into the cobalt blue sky.

So you regroup and trek on upward. And another summit and yet another present themselves, and each time you attain their heights and yet are not rewarded with the peak....

So that when at last you step out on yet another summit that has beguiled you again — but this time there's nothing higher, and nowhere else to go but down — you are distrustful and wary. You glance about furtively. Is this really it? Your whole being has become attuned to serial triumph and disappointment and is ready to resume the climb.

But if then, it really is "it," then you stand perfectly still, lost in the solitary wonder of the enormity of the thing itself, not so much that of your accomplishment.

Now, because it has been such a solitary effort, your emotions are difficult to share adequately with anyone else. It becomes an intensely private celebration, an enigmatic triumph. Though you may have had advice and help along the way, the ascent was ultimately dependent on your faculties alone. Quintessentially it boiled down to a confrontation between you and the Mountain — or more accurately, you versus yourself against the physical world.

If there's a grain of Zen in you, you realize in a burst of insight that "against" should read "with" and thus you are granted the additional blessing of having an authentic tactile experience with the physical and metaphysical world.

Then as you come down and return to base camp, something else wonderful begins to steal over and percolate through you. It's your sense of perspective. It has been enlarged and intensified. You can "see" the whole even though you can no longer see the top. You have been there. You know where it is. And you will never be the same. Your spirit has become as big as the mountain you thought you couldn't climb.

> — J.L.
> August 1992
> Penn State
> University Park, Pennsylvania

Acknowledgments:

Just as mountain-climbing and cabin-building are alike, so too are building a house and writing a book. In fact when someone asked me what I was doing this summer, like an idiot savant I replied: "I'm building a book and writing a cabin." The two acts are strikingly similar; both take a long time and involve a lot of people. Yet in the end, often it is only the author or the builder who gets the most credit. This should not be the case — and especially with this book and this house. During the four-year construction phase, my partners at *This Week* and *The Daily Courier* in Forest City, N.C., had the patience and wisdom to allow me the time out of the office to build my own place. Indeed, they shared in the project vicariously. And as for this book, it was made possible in part by Brevard College and its gifted president, Dr. Billy Greer, whose insight and enthusiasm for this project created time for me to reflect and write the first draft. Penn State gets the credit for enabling me to finish the manuscript by providing computer expertise and equipment, as well as writing time. I owe a debt of unspeakable gratitude to my partners in the dream, Bill and Jeanette Byers, without whom, Hogwild — both the land and the book — would never have happened in the first place. To Maggie Lauterer-Allen, the forewoman of Sweatmore Construction Co., I can only offer this quote: "Some things just can't be said. And they can't be whistled either." But I pucker up anyway. Her foresight started me keeping the journals and her vision was responsible for the house every bit as much as were my sweat and muscles. Without her nurturing and positive presence, I doubt that I could have sustained the effort. Heartfelt thanks to my editorial assistant and graphic artist, Lin Redmond, who has been a true friend and inspiration through the book's production. And finally, a special thanks to the good people at The Appalachian Consortium Press, Cratis Williams' dream. It seems now no coincidence that it took four years both to build the cabin and to get this book to press.

Atop Father Austin's View, Bill and Stuart Byers and I behold a sundown over Hickory Nut Gorge.

1

Beginnings

> *"I'm gonna cash in my hand and pick up on a piece of land*
> *I'm gonna build myself a cabin in the woods*
> *And it's there I'm gonna stay until there comes a day*
> *When this old world starts changing for the good*
> *Now the reason I'm smiling is over on an island*
> *On a hillside in the woods where I belong*
> *I wanta thank Jimmy, Jimmy, John, Nick and Laurie*
> *The No Jets Construction for setting me down a homestead on the farm."*
> — James Taylor
> "Mud Slide Slim," 1971

Saturday, June 8, 1974 — I am 29 years old today. It is time to begin what must be the most singularly ambitious task/dream of my young life: to build my own house.

Maggie has given me a diary, or a journal, a "nothing book," it's called. Inside is Thoreau's quote about stepping to "a different drummer...." Well, that's one tattoo I hear only too well.

She has inscribed the journal thusly: "We are about to embark on an unforgettable journey through summer, one (during) which, knowing you, you'll have recordable thoughts and memories.

"And so to preserve some bit of it, this little book is given to you — to fill with your log of our new Dream. Fill it with sketches, ideas, tales of snakes and sore muscles, anything — just make it yours.

"And when they ask you about writing a book, tell them you've got one bound and printed. By the way, you have my permission to use my name in this one. — with all my love, Maggie."

2 *Hogwild*

Could we make a house out of this barn? In the winter of 1974 we thought so.

Tuesday, June 11, 1974 — The beginning is always the hardest. A book, like a house, opens with an outline, a foundation. (I can see it now.)

So quit dreaming, and just start. See where it leads. Here's the kid who chickened out building the Cub Scout crystal radio set — and I want to build my own house?! Lord knows Jocko's no titan of carpentry, and yet, it does seem feasible.

The Swiss Family Lauterer marooned by choice on 300 acres of Hogwild with dogs, cats and friends; building an environment free of the trash of our cities and towns. Is that so crazy?

No, I answer my doubting side. This is the wisest thing

And this would be the living room. We'd have to preserve the high ceilings too, Maggie decides.

I've ever done. I believe time will prove this to be true.

This all started six month ago — deep in winter when January's gloom seemed impenetrable. Two friends, Bill and Jeanette Byers, had told us of a fabulous tract of land, known as the Whitesides Place, they'd found in the Green Hill community about eight miles west of Rutherfordton up toward Lake Lure. The Byers wanted mainly a glorious old Victorian gingerbreaded house that dominated acres of bottomland down by the Rocky Broad River, and they needed to find friends of the right ilk to come in and help them buy the land.

At first we were simply happy for Bill and Jeanette, that they had found their dream house, but we ourselves entertained no serious notions about joining them... until Maggie made the discovery of a log barn that apparently was very old, and appeared to have been a dwelling in a previous lifetime. There beside it was another structure, a tumbledown two-story frame house that was chock full of lovely wood, heart pine it appears. This was the original Whitesides house, and after the Big House was built in 1915, it became a tenant farmer's house.

I can't say exactly what it was that finally tipped us over the edge; it was a combination of things: our steep house payments, outrageous heating bills, the frustration of trying to keep a 12-room house clean, the desire for more privacy, and urge to be closer to friends like us — and most importantly the influence of my best friend, Mike Thompson, who has just rescued and restored an old place in the country near here. Mike had only marginal experience in carpentry, and yet he just busted in and

Selena and the old Whitesides tenant house, "The Giving House," from which we will salvage a treasure trove of wood.

During one of our first forays to Hogwild — trying to decide whether and how to buy the place: Jeanette and Hannah, Maggie and Selena in the big horse barn.

did this great thing through sheer grit. Well hello, I say, if Mike can do it, why not me? And besides, the basic outline is just "Lincoln Logs."

At any rate, about midway through gray and cruddy January we made up our minds to do it; to sell the big house and throw in with this new wild idea. In fact, the whole scheme was so wild that Bill, in a fit of genius, dubbed the Green Hill property "Hogwild."

And the name has stuck. At first the company of our Earth Family consisted of the Byers (Bill and Jeanette and 3-year-old Hannah), the Kings (Joan and Al and kids), The Walthers (Jim and Rosemary), and Randy Williams, a friend of Bill's from Winston-Salem. But as weeks went by and Bill beat his brains out engineering the land deal by himself, the Kings opted instead for some land they'd found across the river and bowed out.

Tom Cowan, a talented watercolorist, and Father Bill Austin, a so-called "hippie priest" at the black Episcopal church, also came and went, after briefly sharing in the hopes of buying in with us.

When it came down to signing papers and plunking down the money in early February, three families remained, and so the Byers, Walthers and Lauterers signed for the whole place.

I got a Federal Land Bank loan and paid one-third down, $4,500, buying 48.6 acres for $13,500 or about $280 an acre. Imagine just hiking around 300 acres trying to decide which part you wanted. After

A name to fit an audacious idea.

I always thought Bill and Jeanette's Big House looked like a Mississippi river boat left stranded up on the knoll. Built in 1915, it is a wonder of gingerbread and angled walls.

weeks of slogging up and down the mountainsides in mud and rain, we found a secluded valley just east of Bill and Jeanette's Big House; a perfect cove for the cabin.

It had been logged within the last ten years and so we got it cheap. When it came to surveying our tract off, surveyor Bill Hawkins did the work while Jim Walther and I helped to cut the line and hold the pole. It was sort of a dead-reckoning and therefore cheap method of guessing on the map where things were, penciling in a theoretical line and then just heading off in that direction with hopes that the final result would be as close to the 50-acre goal as possible. As it turned out, we didn't do too badly.

We closed on the land in March, and by this time it had become a real Byers family event with Bill's younger brothers, Bruce, Stuart and Burwell, buying in too, along with Randy Williams, Mike Thompson's sister whose name I don't know, and Bill's dad buying the remaining plot originally wanted by Father Austin. (We still call that high point on the land "Father Austin's View.")

Incidentally, why we found this cut-over land attractive in its ugliest bare January state is still a mystery to me. Perhaps we had a vision, and our dreams were so strong.

The land had been logged a couple of years before and was downright ugly in winter. But it had potential — and, it was cheap. Here, we're standing in the Soccer Field looking up at the knoll where the cabin will go.

By April we were working on the land every spare weekend, clearing the site by hand with swing-blades, clippers, and bush-ax. No bulldozers allowed here.

Bill and Jeanette moved into the Big House, and in late April, the younger brothers, both in their early 20s, erected a sure-enough teepee and had set up primitive homesteading in the meadow by the river.

All spring we engaged in the slow and sweaty task of clearing the land. I lost a contact lens to a flying limb on the first day out. We camped on warm early spring nights by the creek, heard deer crashing about in the underbrush, worked on the building site, and played in Silver Creek making dams and terraces.

The little cove that Maggie discovered cuddles up to Silver Creek like a mountain mama. Prior to finding the spot, we'd walked circles around the spot, and it was only after the ridge (Jockey's Ridge) we'd wanted at first proved to be outside Hogwild entirely, that Maggie led me up the cove and then she later found the level knoll where we finally decided to build.

We named our cove Sourwood Holler for the abundance of blooming and early-turning sourwood trees, which have always had something of a mystical quality to me.

"I've got a girl on Sourwood Mountain, hay-dee-ding-dang, doodle-all-day," my Chapel Hill Elementary School music teacher, Mrs. Adelaide McCall, had taught us to sing.

Naming the places has been a fun and frightening task; how many times in your life do you get to name geographical features for all time?

Silver Creek, the bubbling mountain stream that makes something of a peninsula out of the knoll, was named that for a variety of reasons: a poem by Maggie referring to "sweet silver eyes," a Bob Dylan song about the silver singing river, and Thoreau likening water upon land unto the eye of the earth.

The Soccer Field is the long flat meadow below the cabin site and creek, coined by Jeanette for its length and width. Maybe we will play soccer in there someday.

Hogwild, it turns out, was a perfect name for this place. Bill and Jeanette have installed a big yellow mailbox out by the main road with the following painted on its side:

> Jeanette and Wm. J. Byers
>
> are
>
> HOGWILD

Bill and Jeanette's mailbox.

2

How Firm a Foundation

*"A terrace nine stories high begins with a pile of earth;
A journey of a thousand miles starts under one's feet."*

— *Lao Tsu*
6th century BC

Wednesday, June 12, 1974 — After we got the newspaper printed, I hustled out to the land, burned gobs of brush. Health Dept. man came that afternoon and told us where the septic tank should go. Then I contacted the road and septic tank company. They said they'd be here next week.

The cabin site sits down in the cove about 300 feet below the Byers' driveway. We found a decrepit logging trail that will serve as the bed for a road. Once it is dozed and graveled I believe it will work fine, even though it is steep.

Thursday, June 13, 1974 — Sold our house today to the Rogers of Burlington. Good news is that we have until August to move.

At the land we burned more brush, cleared more land, and I chainsawed two big poplars on the cabin site that had to go. Too bad, because I'm trying to hang on to every tree I can, including one bent big dogwood that was bowed and practically destroyed by a partially fallen poplar.

Friday, June 14, 1974 — We have broken a garden at the land just across Silver Creek in the Soccer Field. Much to our amazement we found several Indian pottery shards in the loamy soil, confirming Bill's supposition that this area was once populated by Cherokees, who either lived right here or used the nearby river as a trading path.

After gardening in the sultry heat, we knocked off and just went and

The log barn sits up on Hogwild Blvd. 100 yards above the building site. The tenant house is in the background.

sat cooling off in a little waterfall way up Silver Creek above the cabin site.

Down by the creek we have cleared sort of a play area, and I've built a rock fireplace. Tonight we had a great fire and did roast corn and burnt wieners. Bill and Jeanette came by with Hannah and we sat up until dark talking by the firelight about our dreams.

Saturday, June 15, 1974 — Borrowed a truck today and drove up to Jonas Ridge where Dr. Austin Hyde has a cabin. He has donated a bunch of wood he's salvaged, fine old 14-foot beams and tons of siding.

Also, we bought a four-man tent so we can camp down in Sourwood Holler this summer while building.

Sunday, June 16, 1974 — It was Father's Day, so we had Maggie's dad down for a picnic on the land, but after carrying the old picnic table down to the creek and pitching the tent, the heavens opened and it came a frogstrangler. We had to retreat back to the Byers' Big House for lunch.

Later, we worked on labeling the logs. I've devised a method that goes like this: file cards are marked with the log's identification number, wrapped in "Baggies" and stapled on either end so I'll know original positions. Also, I've photographed the barn extensively for record-keeping.

The front of the barn I've labeled A and B, 1-13 from the ground up.

The back is labeled M (for Maggie) 1-13; the north side is labeled W (for Wild) 1-13; and the south side X (because I couldn't think of anything else cute) 1-13.

The barn's south wing is supported by two huge horizontal logs: Big Mama and Big Daddy. Pop's on top. Then the vertical support beams are Zonk 1 through 9, so named after one of my stranger nicknames. The north wing's beams are less intact, Sad Sack being the name of one of the partially rotted sill beams, and the verticals are Skipper 1 through 9, because that's one of Selena's nicknames and that's her bedroom to be.

By dusk Maggie and I had finished labeling the cabin. (I notice I have started calling the barn the "cabin," and that's projecting quite a bit, because what we're talking about is surely only a barn -- a low entrance, second story hay door, dirt floor and stalls.)

There are two rafters about eight logs up that span the center room. Maggie and I crawled up there as the evening settled in. We just sat there dreaming and listening to rampant whippoorwills going Hogwild.

Oh yes, also unloaded the beams and siding from Dr. Hyde's cabin in Jonas Ridge; turns out the wood is solid chestnut! Couldn't be more tickled.

Monday, June 17, 1974 — After a hard day on the paper, we got out to the land by 6 p.m. for a final day of cutting and burning in anticipation for the road and septic tank folks due here by Wednesday.

Stopped well after sundown and went by Bill and Jeanette's for tea and sympathy. Bill and I got into an endless planning/strategy session about how to take the barn down and move it. All appears ready to go into action. I should be able to start the foundation (we hope) by Thursday, June 20...take say 10 days for that, and possibly be moving logs down and out by July 1. Hope against hope.

Wednesday, June 19, 1974 — Today was supposed to have been D-Day for the road and septic tank, but they failed to show. That threw me off. Guess we've got to get used to that.

I discovered too that the log labeling done with Bic "Banana" felt-tipped pen has washed off in the rain. Even wrapped in Baggies, the ink dissolved. Will do it over in something wax-based like Crayola Crayons.

Thursday, June 20, 1974 — I got to the land around noon and began working alone salvaging wallboards from the tenant house. It's beautiful wood — wide tongue and groove pine without any knots. I found a lovely fireplace with huge hearthstones. The house is just full of this

How Firm a Foundation 11

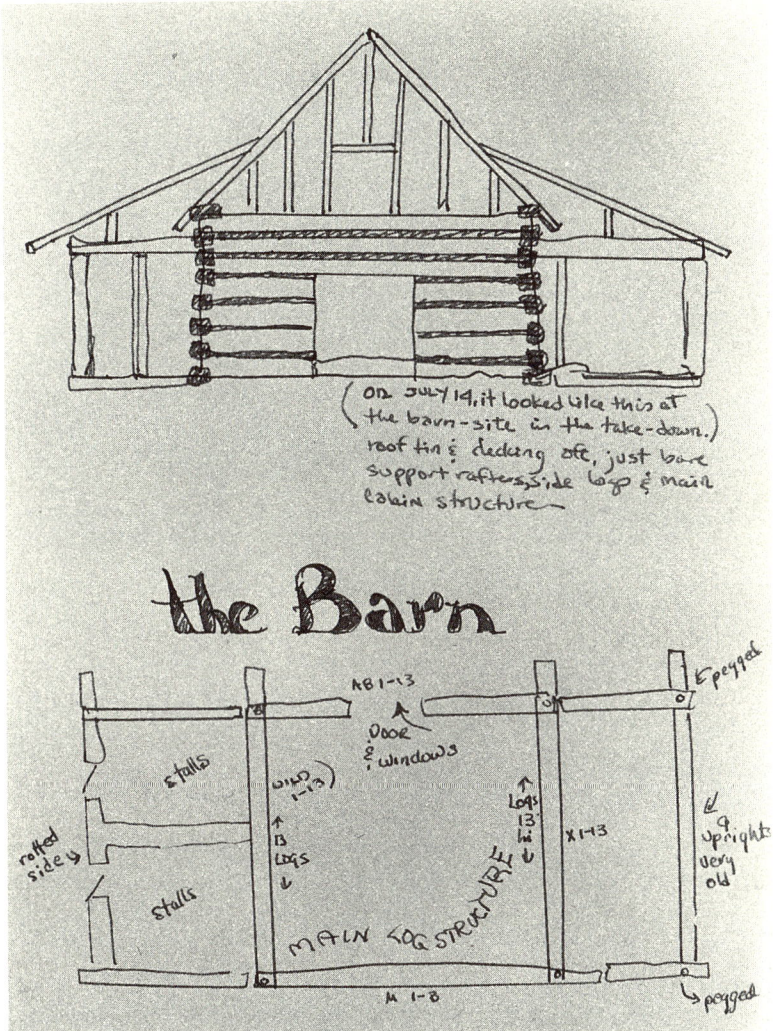

Front and overhead views of the barn structure.

heart pine. We'll use it all. Got more than enough of the faded whitewash pine to do the kitchen. Worked all morning alone, the dogs and the hearth fire as company.

A huge rainstorm blew in that afternoon. The old house groaned and clapped tin roof sheets in the high wind. It was truly wild.

Mid-afternoon a concrete truck arrived and brought 200 blocks and six bags of Brixment for the foundation. But of course I don't have a road down in the Holler yet, so the entire load had to be left beside the road at the top of the cove; reckon I'll have to tote them in by hand...

12 Hogwild

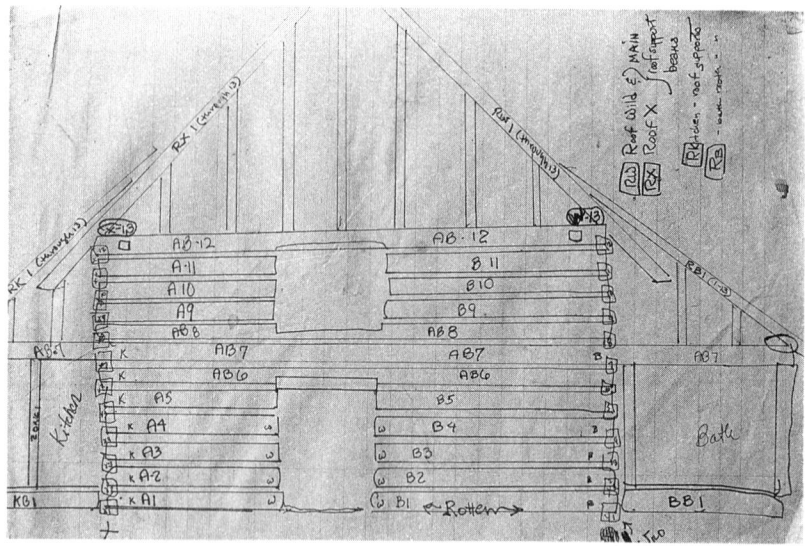

The barn labeling system proved its value with time.

Maggie came later that afternoon and together we attacked the upstairs bedroom, salvaging a ceiling that was water-damaged, plus one side wall. It was dirty, hot and rewarding work; like getting something for nothing.

A summer storm blows across Hogwild as I salvage in the tenant house and light a fire. Ruby the Weather Dog seeks reassurance.

While pulling back one wallboard, I uncovered the lair of a big coiled blacksnake which jumped out at me as if spring-loaded. If there's not a track event called the "standing backwards leap," then maybe there ought to be, and I'd have qualified. I leapt backwards and fell over a pile of wood. Along with the horde of yellowjackets we liberated, it was a pretty exciting encounter.

We washed up in Silver Creek, cooked hoop cheeseburgers over the house's ancient fireplace, having a wonderful evening camping in the old house.

Friday, June 21, 1974 — Today is

Saint Han's Day, the Midsummer Night's Eve. Bill tells me that traditionally during this, the shortest night of the year, the old Scandinavians partied like wildmen around a bonfire. So that's what we're going to do.

I worked all day digging foundation ditches for the footing. About halfway through, Bruce Byers brought me a gas-powered ditchdigger from North State Gas Co. That is some monster. It was all I could do just to keep the thing under control; like trying to hold back a raging bull. But by the end of the day, I had completed the basic outline of the cabin: 18'x22' and not squared at all.

Wednesday, June 26, 1974 — While trying to place a sturdy log house down on the sloping knoll in the Holler, we're aiming at maintaining the exact atmosphere of the place. I want to leave as many trees as possible. If I dig the foundation footings by hand, there will be no need for a dozer, whose maneuvering would no doubt destroy trees needlessly.

The idea is for the cabin to look as if it just grew there, truffle-like. It should appear as if the cove gently cradles around the old log cabin. In many ways, I'd love just to have a path winding down through the woods to the cabin. But I'm having to compromise here. I need a road to get the logs, blocks and rocks in, and today it was the septic tank that had to get in.

Today was in/credible — as in "beyond belief."

Maggie in the "kitchen" wing with logs labeled Big Daddy overhead, Big Mama below, and vertical Zonk uprights.

Barefoot Selena clambers around in what will be the living room.

I got there at 9 a.m. and the septic tank folk were there chomping at the bit to be done with it. This was the same outfit that was supposed to do the road, but that part of the operation seemed absent. Because I'd never had any experience with this sort of thing, I just kept expecting them to know what they were doing and do it right.

How Firm a Foundation 15

The barn on Hogwild Blvd., logs labeled and ready to be moved down into Sourwood Holler.

It was a collision between ideological opposites, between a young pioneer wanting to preserve his habitat and mad-dog industrial widget-construction mentality. The job superintendent, a Rod Steiger look-alike, growled at me right off: "Y'orta hadda bulldozer in here," he said as he huffed past me up the little logging trail.

Fantastic. I wagged my head, dreading the next hours. And I had plenty to dread. These guys just didn't respect the Land. Trees meant obstacles to these men. Knock 'em down if they're in the way.

I felt like a guest at my own place.

Saving trees became my main concern as the contractor seemed to square off; he just wanted to get done and get out. I wanted it done right and with minimal damage. But the cabin's remote location and difficult terrain really hacked him off. Here was the line-up: The septic tank like a big gray coffin riding on an eight-ton, E-flat, double-clutching eight-wheel Mack truck as long as the log barn, a platoon of dump trucks with gravel, and a back-hoe to dig the hole and the aerating field below the tank...all this was backed up Bill's road, their diesel engines growling and ready.

When it all started, there was no stopping the debacle; there could only be damage control on my part. Once, when I was pulling back an oak sapling around one of those immense dump trucks as it eased down the woodland path, the driver looked out at me and grumbled scornfully, "Getta ax."

Horrified by the total lack of understanding and hanging in mid-air from this dude's monster truck, I had the presence of mind to paraphrase good old John Muir: "Any fool can cut down a tree," I responded. "But it's just a little bit tougher to try and save one."

That fell on deaf ears, for the driver looked at me as if I was the goldangest fool he'd ever seen. So there I was, dashing around pleading with machine operators to watch that sourwood please, and couldn't we just lift that dogwood over the bed of your gravel truck, thank you...

The superintendent stormed past me about mid-day, and snorted, "If I'd seen this job, I'd never come."

Slowly it began to dawn on me: they were going to get all those trucks down here, get that septic tank in, get stuck up to their proverbial axles, and then put the road in.

I watched with a mixture of fascination and horror as every one of those trucks that rumbled down the steep, narrow logging path got hopelessly mired in the muck they themselves had made out of my humble entrance.

So, to get those trucks out, the backhoe had to spend much of the morning pulling trucks inch by inch out of the cove, and the result was two-foot-deep tractor tire gouges in the "road." Plus, the fact that they busted two logging chains in the effort didn't improve the workers' temperament.

When things quieted down, I had a chance to get on with my work of starting to lay the concrete footing for the foundation. The workers watched in undisguised amazement the primitive foundation taking shape.

"You dug this by hand?" one asked, just to make sure.

By the end of the day, the septic tank was safely snuggled in its bed, the gravel field installed and buried nearby, I had five feet of footing poured, and the "road" looked like dark brown mashed potatoes.

By this time I was fully prepared to agree with one trucker who offered kindly, "You'd be better off just to go in the bushes."

It had been a day of sound and fury, of Man demonstrating his power to thrash the Land into submission. And for me, the encounter confirmed my worst suspicions, reinforcing my conviction that I'd do this one by myself and do it right.

When they had all left, when all the clanking earth-moving, eating and chomping equipment had rumbled out of my valley at dusk, I walked to the brink of the cove — and something seemed to take hold

How Firm a Foundation 17

of me; I froze in my tracks, barely breathing.

It was the sound of the woodland silence.

Holding my breath not to mar that spell, I opened all senses of sight, smell, hearing, feel, and yes, taste to the land, as it seemed to rush back protectively from all sides to reclaim the little cove in a wilderness hug of green utter silence, as if to comfort my cabin site for the day's injustices and brush away sourwood tears.

And right then I promised my cove I'd never again let Man inflict himself upon her fragile beauty; that it'll be done the hard way, the right way, and in such a way as not to disturb a single leaf. It's a promise I intend to keep.

Maggie returned from a late-night jazz singing practice and we sloshed down to the campsite in the rain, mud and muck. We both just passed out and slept great on our cots amid the staccato lullaby of fat raindrops dive-bombing off the overhanging boughs and slapping flatly on the canvas tent.

Thursday, June 27, 1974 — We had a fine breakfast up at the tenant house because there's plenty of dry wood there. This morning the sun is shining but all the wood down at the campsite is soaked.

My initial footing of yesterday was dry, though. I spent the morning putting in the remaining footing on the entire north face of the foundation. I'm ready to lay block when it dries. Happy day.

By 2 p.m. the rains returned, and a real downpour too. So I retreated to Jean's Kitchen, the local country cafe in Green Hill, and wrote herein and had lunch. The rain seemed to want to settle in, so I returned to salvage in the tenant house. Maggie returned from shopping and errands, and we spent four hours taking out nails, stacking boards, and uncovering two more fireplaces. By sundown we had three fires going, and it was cozy even

Doing it the hard way: toting in blocks one by one and digging footings by hand.

18   *Hogwild*

I've laid the first course, and momentarily lose control.

though the rain continued unabated. We cooked supper there over the fire, and then worked by firelight salvaging until around 9 p.m. we drove down to B and J's for real Hogwild blackberry sundaes, and watched Kung Fu, a TV program Bill likes for its Eastern philosophical approach. And then we took our flashlights and stumbled off into the woods to our campsite and so to bed.

Friday, June 28, 1974 — When the biblical folk wrote of building your house upon a rock, and the hymnist penned, "How Firm a Foundation," they knew the value of building a house, however humble, upon something more substantial than what many of our ticky-tack subdivision tract houses rest upon.

   This log cabin venture took another historic first step this morning with the laying of the cornerstone, the first block. This minor miracle occurred without official fanfare, mayoral ribbon-cutting, or photographers snapping pictures of silver shovel-wielding investors. Just me in my grubby red gym shorts, kneeling in the mud. I placed the first block snugly in its bed of concrete to the sibilant hurrah of Silver Creek.

   And I stopped to look about if the forest or my cove was paying proper attention to the momentous event. No, there was no change in the land. Just a hole with some hardened concrete. But otherwise the wildness remained unimpressed.

   Well, I was impressed. I've had absolutely no previous experience in block laying or any other form of construction for that matter. The city boy turned country man discovers at every turn that there is so much to be learned. So much that I don't know. And I learn painfully, trial and error.

A foundation, for instance, requires engineering comprehension I'd never dreamed of — especially when building on a slope. The first-time-out builder learns (1) start it level and keep it that way, and (2) the higher the ground the deeper you dig, and the lower, the shallower you dig.

By mid-afternoon I'd put in 14 blocks (all carried down the Holler by hand, two at a time), the entire first row of the north face — all perfectly level if a little crooked in places.

Lesson # 47: No one told me to keep my hands out of the concrete. After a couple of days of handling the stuff, my hands look like those of Dustin Hoffman playing a 115-year-old in *Little Big Man*.

So knotted with new muscles and calluses, my hands feel like those of pioneers and they do vast injustice to guitar strings. Reckon the real old-timers playing by the light of lamplight at day's end got used to playing with those strong, gnarled tools, the hands of a man of the land.

Saturday, June 29, 1974 — Observing my work, Jeanette says, "It actually looks like something's being built down here now." Today was a really good work day — completed almost the entire north and west faces, three to four blocks high on the foundation. All this goes very slowly, since we're still without a road and forced to haul everything in like coolies by hand: blocks from the top of the cove two at a time, buckets of sand that we've filched from Broad River sand piles at the swimming hole three miles away, and water toted by bucket up from Silver Creek — and all is mixed with an old broken-handled hoe in a borrowed mortar pan.

Maggie went to sing at Land Harbors in Linville tonight. Selena and I camped out at Silver Creek site.

Sunday, June 30, 1974 — Maggie didn't get back from the gig until 3 this afternoon. No phone down here so she couldn't reach me. I was really worried, and contemplated for the first time what life might be like without her: unimaginable and empty. Selena had played all day down at Hannah's, and I worked alone laying block, and even though it was a lovely day down in the Holler, I was strangely lonely for the first time.

When Maggie returned in the mid-afternoon I was so relieved I was ashamed for needing her so much. She had finished singing late and had opted to spend the night at her mom's in nearby Crossnore.

We concluded the day with a Byers family picnic down at the Big House, and if there was a celebratory feeling about the affair, maybe I was imagining it; or maybe I was just privately rejoicing for the return of my raven-haired mountain woman.

Monday, July 1, 1974 — A newspapering day for me, and the road is still in shambles from the Great Septic Tank Massacre.

Wednesday, July 3, 1974 — Glory be! The wilderness road happened! As fate would have it, no sooner did I berate the drivers of Big Things ravaging Sourwood Holler than I have to eat those words. I have met a backhoe operator and a bunch of dump truck operators who are virtuosos at their craft.

Roger Watson, the backhoe operator, is a math teacher at Cool Springs Junior High during the school year. On the backhoe he is a calm, meticulous artist gently at work.

I could practically see Roger's mental computations and mathematical exactitude translated to backhoe tractor blade as he went over and over the road until it satisfied his painstaking specifications. Not a single tree felt even the vaguest whisper of his blade.

And the dump truck fellows delivering the gravel this time did a magnificent job of backing in part-way and carefully getting out again.

Indeed, one dude tickled me... after dumping his load, he said, "Fellers back at the office have been talking about this place." He grinned at my wilderness road and chuckled.

This time, I was only too glad to chuckle along with him.

Thursday, July 4, 1974 — We traveled to Snow Camp to sing for the dedication service of the new outdoor drama, *The Sword of Peace*, which is fitting, because it's about the North Carolina Quakers during the Revolutionary War -- and my ancestors on the Rush side of the family were Quakers in this state and very much a part of that history.

Friday, July 5, 1974 — A sultry thing, July. Air conditioner busted in the front office and it's so hot up there you can't think. I'm writing in the air-conditioned darkroom with my trusty black Royal nestled on the easel of the enlarger.

Today I laid several more blocks and more footing around the northeast and southwest corners after re-measuring and relocating strings and stakes.

Maybe it's because I'm out in it all the time now...but it seems like this is one wet summer. Every time I crank up to lay block the heavens open and dump on me. Still, I'm three-quarters of the way through the foundation.

Saturday, July 6, 1974 — Raining buckets again. Drove the Ghia down the road loaded with concrete blocks (Ghias aren't made for such yeoman service) and the road is so mushy yet, the car couldn't climb out. Precipitation seems adamant about flooding out my best efforts. I really wanted to finish the foundation this weekend.

I'm learning about construction deadlines vs. the fickle elements, and about reconciling the two.

Perhaps wilderness road will pack down and firm up with time. Otherwise I'll be forced to rename it "Four Wheel Drive."

Today I laid the corners on both NE and SW sides and am ready to charge on towards the final SE corner — if only the weather will cooperate.

That afternoon some real friends came to my rescue. My old best buddy from Chapel Hill and early days at *This Week*, Danielle Withrow, blew into Hogwild accompanied by the Chigger-Slappin' Shine-Sippin' Summer Jug Band of Goldsboro (where Danielle now works as director of the Arts Council).

What a crew: Dan Mason, Alice Creech and Rick Ervin all helped by toting blocks down to the site. And of course, an outfit with a name like that couldn't work without a marching tune, so they accompanied themselves on kazoos, playing "Colonel Bogey's March" (the theme music to *The Bridge Over the River Kwai*) as they came striding into the clearing with their blocks.

What a great afternoon; I got a lot done on the foundation. And that night on the campfire we cooked up a goulash sort of mixture we call "Union Grove," named for the fiddler's convention where we met. The dish consists of anything and everything anybody has to contribute.

After supper we lit into

Kazoo'ing while they work, Rick Irving and Dan Mason carry blocks.

The Chigger-Slappin', Shine-Sippin' Summer Jug Band comes to our aid.

jug band music and on through the night and then when the monsoons came, we piled into the tent and kept at it (all nine of us) while outside the campfire mysteriously kept blazing through the downpour...until it got so bad that the tent started leaking and we had to hightail it down to Bill and Jeanette's and just passed out on the floor and slept fine.

Sunday, July 7, 1974 — We returned to the campsite for breakfast, and whipped up a gobbledygook omelette — the eggy counterpart to "Union Grove." The sun came out and I was able to lay more block. That afternoon we all went swimming in the Rocky Broad, and cooked supper at B&J's. Long live the Chigger-Slappin', Shine-Sippin' Summer Jug Band! They really made a difference.

Wednesday, July 10, 1974 — Nearing completion of foundation. Good weather allowed me to work all afternoon after getting the paper out.

Thursday, July 11, 1974 — A really good work day. I nearly completed the blessed foundation. Just 15 more blocks to go and I've used up all 200. So I had to go to Henson Timber and buy the remainder.
    Lewis Barnes, who is sort of my spiritual mentor there, kids me unmercifully whenever I go in and ask inevitably dumb questions. And this time he got a hoot out of seeing me loading 15 concrete blocks in the tiny Karman Ghia.

Friday July 12, 1974 — Today I completed the 18'x22' foundation. I couldn't be more happy about it. The damned thing actually came out level!

I christened the foundation with a Rolling Rock, being fresh out of Dom Perignon at the time. And then I christened myself with a brew as Maggie took pictures of the laying of the historic Last Block.

Incidentally, to save myself from having to tote out garbage, I've been stashing my empties in the air spaces in the concrete blocks. I reckon in the 24th century, some archeologist will unearth the cabin and find a rich lode of historically precious beer bottles buried like time capsules from 1974.

At last, the foundation is completed. We pose for ceremonial photos.

3

The Summer of '74: Barn-Razing and Cabin-Raising

> *"No great journey was ever begun with apologies"*
> — *Chinese saying*

> *"Well begun is half done."*
> — *Aristotle*

Sunday, July 14, 1974 — We got up early, having spent the night again at the campsite. Along with Bill and his cousin Terry Klavan, an engineer from Charlotte, I attacked the roof of the barn before it got too hot. It was one of those days where you know in the early morning that

On a morning as hot as blazes, Bill, cousin Terry Klavan, and I attack the roof.

it's going to be a baker, and it would make the tin roof a veritable frying pan for would-be woodbutchers.

We clambered around like kids on a jungle gym; it was admittedly scary but exhilarating, taking the old tin roofing off and slinging the sheets below, where Maggie and Jeanette stacked them. Next we took down the old rough-sawn pine roof boards.

How tall the thing seems when you're 30 feet off the ground working on a 45-degree pitch and you're throwing away your own support system.

The sun blasted down making me wish out loud I had my Caterpillar cap ("Cat Hat," I call it), causing Bill to opine that if I did, I'd be a..."Cat on a Hot Tin Roof."

I guess we were getting altitude sickness. We got pretty silly; the three musketeers slinging wood and tin about from a giddy height. We finished around noon and were hot, sweaty, and grimy from old barn grit and dirt. We decided to call it a day, and all trooped down to the swimming hole and threw ourselves in the cooling Broad as a reward.

The result of our work is that the building now looks less like a barn and more like a cabin. Now I am ready for the final barn razing itself.

Jeanette hits the water down at the old swimming hole after a day's hot work.

Wednesday, July 17, 1974 — Today has been absolutely unreal: Bill and I, just two men, took down the entire barn and we did it completely by hand. Starting to work at dawn, we tackled the problem of how to remove the roof rafters. This was no small undertaking, for the pitch was easily 45 degrees and the rafters met at the ridgepost fully two feet above our reach. We had no idea how to take down the rafters and still keep them intact.

With careful rope work, the barn comes down, log by log.

"Well, let's not think about it too much," Bill said. "Let's just get up there and start doing it."

And so that's what we did. We just did it. Starting at the outside gables, we unhooked RB and RK series (Roof Bath and Roof Kitchen) where they joined the top sill log, number 13 on either side and nicknamed "Big Deal" (for one of the Jets from *West Side Story*) and so named because of their immensity.

As we went along, we learned by doing, and figured out that the process of unhooking every other rafter (so we still had something to hold onto), letting one end drop in a lazy arc, would loosen it at the other end where it met that out-of-reach ridgepole. Then on the ground, we could simply wrench it free.

While Bill shimmied down one Big Deal and loosened rafters on his side, I did likewise on the opposing Big Deal.

Through this process, with log rafters falling all about us, we laughingly developed a sort of a battle-cry for the day...

"You OK?" I'd holler after a series bit the dust.

"Yeah, I'm OK," Bill'd call back.

Naturally this led us to conclude slaphappily that the title of a book on how to survive a barn razing/cabin raising would have to be titled I'm OK, You're OK.

At last, we got it down to three series of rafters on each side. Then, we saw he could unhook that skeleton at the middle, and simply push either end down in a controlled fall. So that's what we tried — and it worked.

Amazingly, none of the skinned pine pole rafters broke. And we were left standing there 15 feet up on our respective Big Deals, now the tallest thing around. It was a heady, verily experience.

By midday we were ready for some heavy equipment. I'd contracted with a young man to come with his tractor and forklift, and use this device to remove the logs, drag them down to the Holler and then put them back up, again using the forklift. It seemed like the logical way.

But it was not meant to happen that way. The dude arrived on the forklift and his friend came on the tractor, but once they saw the size of the project, they both started jawing about, "Hey, man, I mean wow, this sure looks like a lotta' work, man..."

While they stood around flapping their jaws, Bill and I devised a method for removing the logs simply by means of rope and muscle. See, we were on a builder's high, and weren't about to let indecision stop us.

We found we could rope each end of the top log, and lower one end at a time, using the next log down as a brake. We began by unpegging the top sills, the Big Deals, and we must have lowered eight logs while these two helpless guys watched.

Once, while I was lowering my end of a log, a rusty nail bit into my stomach; I didn't even notice.

"Hey, man, you're bleeding," said one of the groundlings. They didn't understand, we were having too much fun to hurt. Finally, the pair of shrinking violets belatedly decided to help.

But they were pathetic; they dragged two logs down to the foundation and returned to tell us that was all they could take. I couldn't believe it. I had contracted to pay as much as $250 for their work.

But it worked out for the best. Bill and I had discovered the Eureka Factor, and we were doing it all ourselves for free and having a ball at the same time. Filled with confidence in our new-found strength, we

Bill, Stuart, and Burwell Byers pitch in, and it's done.

lowered logs until, around 4 p.m., we got to the seventh round. Then brothers Burwell and Stuart arrived and pitched in. We experienced a rare 30 minutes in which we took down the last 28 logs using what we called the Four Corners Offense (with apologies to Dean Smith).

Using our rope method, each of us took one top corner of the barn, and alternating as our turn came, we gently lowered our corner to the ground, until the barn's logs lay scattered about our feet, and we cheered.

Incredible. We had met adversity and overcome it with sheer common sense. A keystone building philosophy has been uncovered. I'm reminded that it's something I've heard before but hadn't fully internalized: when Maggie was asking 90-year-old Aunt Nettie of nearby Bills Creek how she learned to quilt, the wonderful old woman just smiled and told us, "Why child, y' just *do* it!"

What Bill and I have discovered is that if you just begin, then the answers have a way of revealing themselves to you. The process is its own solution. But the builder has to be attuned to this possibility for novelty. And you have to give yourself permission to go out on this "leap of faith," as my friend Reggie Cooke calls it.

I may never be the same again.

And the beautiful thing about all this is that now it seems equally feasible to put the process in reverse — and reconstruct the entire cabin using ropes and the same Four Corners Offense. Only, this time, gravity won't be on our side.

At the end of this zoomy day, we four guys went swimming in the Broad to wash off, and that was so fine. It helped, as my legs are cramping terribly tonight. But it's worth the price. I was a monkey-man on the Lincoln Logs.

We all went out to supper and returned to bed early at camp as Steve Walker had been contracted to come early with his tractor and haul logs. Also, Bill's friends Charlie Ratcliffe and Cousin Gene Ham are assembling so that we'll have a slew of folk here for a cabin-raising!

Somebody pinch me.

Thursday, July 18, 1974 — Another fabulous workday. The whirlwind roared on. Steve Walker with his tractor arrived at midmorning but had little success dragging logs with the cable he brought. So I hightailed it to town and bought one monster of a logging chain, and raced back to the woods.

Hooking up the first three logs, Steve sounded like a country Eeyore: "You'll be the luckiest man alive if you get all these logs down

"Crazy Steve" drags the logs from the old site to the new foundation in a day.

there by today," he said gloomily — and then proceeded to amaze himself and us as well.

The thing worked, and the pace picked up dramatically. By midafternoon the pessimist was converted; he'd done it and done it all. I paid him $75 after much talk about doom and gloom prompting me to call him Crazy Steve. It was all very depressing and sobering but worth listening to.

"People will be killing each other for food within six months," Crazy Steve said, advising, "Keep a gun, and better save your bullets — for people!"

Later that day, we start in, and get four courses up.

30 Hogwild

After the prophet drove away, Burwell and Stuart agreed to help us lay the first four series of logs. This was the most exciting moment: actually placing the big sill log, X-1 and W-1 in place on the foundation and having everything fit. Have I mentioned this before? But the foundation rectangle wasn't square, and so that the notches would fit correctly, I had to basically build the foundation "wrong," that is to say: true to the fit of the old barn, even though it was whackeyjawed.

Working on, the three of us put up 12 logs, or three series, and past there, we saw we'd need fresh recruits before going any higher. The logs we were hefting were 22 feet long and 10 inches wide and 14 inches thick. That's a lot of log.

Again the gentle Broad River accepted our filthy but honestly dirty bodies with its soul-reviving whirlpool bath. Afterward we all supped together at Bill and Jeanette's. Gene Ham, an old friend of Bill's from Sewanee, was there with friend Kristin along with my good man Charlie Ratcliffe (whose services I have acquired to help me roof) and wife Susan. We had a fine veggie supper and talked of the D-Day to come.

Friday, July 19, 1974 — Cabin-raising! A day I'll never forget. It began with the four of us, me, Bill, Gene and Charlie, going at the logs with strong backs and lots of laughing. We did it by pure heft, and using the labeling system we'd devised. The file cards (marked with Crayola Crayons, wrapped in Baggies, and stapled on either end of the log) had

Here goes: Bill, me, Cousin Gene Ham, and Charlie Ratcliffe start in with the lower logs.

The logs go up with relative ease until they begin getting over our heads...

...then it takes everything you've got.

survived being dragged down into the Holler.

It was sort of like building a house by Mattel; we just followed my diagrams and hunted for the right log in the piles. Whoever found it would holler out "Wild-8!" or whatever, and then we'd get together and tote it over and heave it up into its place.

By noon the cabin was rising about us seemingly of its own will. But we had gone about as high as we could without busting a gut. The women up at the Big House had fixed a hearty lunch for the pioneering menfolk: chicken gumbo with veggies from our gardens, and Kristin's "health bread." It was great, and we rested sore backs until Burwell and Stuart arrived with friend Chris Riegert

32   *Hogwild*

Gene Ham and Charlie Ratcliffe work one corner.

The two 40-foot monsters that span the wings challenge us, especially on the deep side.

around 1 p.m.

Our work force was now seven, and they were magnificent. The higher the logs had to go, of course, the tougher the work became. The guys pulling the ropes became airborne like church bell ringers, at times completely off the ground with bodies perpendicular to the inside of the log wall.

On the outside, there were three guys hefting each end, and up on top were another two pulling the rope from that position. There were times when each one of us was strained to his physical limits.

Up she goes, one corner at a time, and with rope handlers on the inside walls.

The higher we go, the more the building process reminds us of playing on a jungle gym.

Looking back on it now, I realize that if anything had slipped, or anybody had fallen, it could have been taps right then and there. The massive heart pine logs would have mashed us like buckwheat cakes. But we were working too hard and having too much fun to think of anything like that then.

And we were being swept up in the momentum of the building; I guess it's the spirit of a cabin-raising. The closer we got to the top and to completion, the more the excitement mounted. Maggie added to the sense of history being made by running around taking pictures of the whole process.

When about midafternoon it started raining, no one seemed to notice. The light summer shower came as sort of a summer elixir sent to soothe sweaty workers.

We heaved, grunted, and shoved on through the drizzle, until at last, the two behemoth logs, Big Deals X-13 and W-13, were manhandled into place.

Then came the finishing touches of lining up the biggies and actually pegging the original pegs back into their old positions. Perched high atop one corner, I whacked the final peg home and cheered, waving my hand-ax like a Hogwild thing.

We clambered down the logs, grinning broadly, slapping backs and popping beers. What a cosmic rush to see my house, as Charlie put it, "flipping up before my very eyes."

I am boggled; in the space of three days friends and I had taken down, moved and rebuilt the core of the house. This barn-razing/cabin-raising has been a total experience. I've never worked so hard, for so long, for such a stake. Nor have I ever been so dirty, so happy and so exhausted in the space of three decades since I boarded this spaceship earth.

I know this is just the beginning in many ways. Floor, roof, wiring, plumbing, windows and the Ten Thousand Little Things have to be done before it's finished. I'm confident we can do it all in good time. But the first and most crucial step was the log-raising.

It was just as good as I hoped it'd be; a once-in-a-lifetime experience.

Barn-Razing and Cabin-Raising 35

Then at last, there it is, 13 courses high, standing proud, solid, and straight.

4

Roofing: In Which We Become Acquainted with 5,500 Shingles

"Gimme Shelter"
  — the Rolling Stones

Sunday, July 21, 1974 — As old Aunt Nettie says, "Good friends is good friends." And today more good folk came to help. After the Organic Garden Club picnic down at B&J's, several adventurous high-wire artists joined me for the aerial act. Crazy Steve was great up there, along with Hogwilder Randy Williams.

First, we devised a sort of a poor-man's scaffolding out over the core of the cabin (25 feet straight down, and no net). Then we put up the kingpost — that's the center roof brace, and then when we tried the corner rafters, we discovered the top of the roof extended two feet out of reach.

Necessity being the mother of invention, the only alternative was to whack the kingpost down to within reach , and therefore, lower the entire roofline by two feet. Aesthetically I stewed over this, but what to do? Friends were there to work, and, as friend Jim Dunn says, this roadblock to our progress was being "obstacular."

So we did it, and work continued. Meanwhile, another item of aesthetic importance solved itself. That was the roofline: I wanted it to have the swag that the barn had. Sounds crazy maybe, but I liked that gentle swayback that comes with age; I had no idea how it happened or how to reproduce it faithfully — until we got into it.

Again, the Eureka Factor kicked in. We built the ridgepole out of one-by's nailed together (because we didn't have a 2x4 30 feet long) so that it was supple, and long enough to accommodate the overhang I wanted: not much on the chimney side, and about five feet on the creekside. The idea was to give it the flare of a Swiss chalet; sort of a prow of a roof.

Getting up the basic ridgepole and ridgeline takes more doing that I imagined. The Organic Garden Club pitches in.

Well, when we put the ridgepole up, it naturally sagged slightly because of its light construction, and I could see from that suggestive framing, made really by a mere seven pieces of wood, the bare outlines of a complete roof. It was like a sketch done in wood. And the swag seemed to be there of its own accord. Eureka encore!

Thursday Aug. 1, 1974 — Notes to myself:
— Move (somewhere) by Aug. 15 at the latest. Find a house.
— Complete foundation blocking on bathroom and kitchen wings so roof can start.
— Get 2x10s and fill in for rotten logs.
— Notch roof rafters — by hand, in cooler time of day.

Monday, Aug. 5, 1974 — We have power! The temporary service box has been nailed up to the big poplar outside the future kitchen, and Rutherford Electric Membership Corporation sent a team out here to connect the electricity. I abhor the powerline coming down through the woods; it's offensive, and when all is done, I'll have the powerline buried.

One of the linemen looked at what I was doing and asked me with honest amazement: "You mean, you're gonna live here?"

Thursday, Aug. 8, 1974 — Today and yesterday Charlie and I got all the rafter poles up on the main section of the cabin. And yes! The swayback is reproduced in the roof because — of all things — the top sill logs (Big Deal X-13 and Wild-13) were laid sideways and have been like that for gosh knows how many years — so the natural effect is for a sideways log to sag in the middle from its own weight, whereas a log laid upright keeps its rigidity.

Friday, Aug. 9, 1974 — Amazing. Simply amazing. Today America got a new president.
　　Charlie Ratcliffe and I were working on the bathroom wing joisting and framing when Jeanette hollered at us all the way from the Big House: "Nixon's quitting! It's on TV! Come see!"
　　We dropped our hammers and ran out of the cove, up the hill, and down the driveway to B&J's, talking excitedly. Can this be real? Have they finally gotten Tricky Dicky?
　　Hot and sweaty and clad in cut-offs, dusty nail aprons and brogans, Charlie and I sat stunned in front of the TV. We knew we were watching history. Serves the old crook right.
　　While Watergate has been brewing and Nixon's administration coming unglued in this summer of '74, I'm proud to say I have been building something of honest substance out here in the woods, far away from all that subterfuge and political chicanery — something that's going to outlast many administrations to come, too.
　　It's strange to be building a dream house, and yet not be able to move into it. We've got to get out of the Pea Ridge house by next week, and so we've been looking for temporary shelter while we finish the cabin.
　　This week Jim Walther put us on to an old abandoned house next door to him in Rutherfordton. If the town had a condemnation ordinance, this place would have been used for practice by the fire department. Jim and I scarfed it up for $7,500. It's a steal, but it needs a lot of work.
　　Built around the turn of the century, it has no insulation, the windows are the original drafty big-paned jobs, it has no subfloor — and what's there is old pine, painted brown. (Can you believe they painted the floor?) The wiring and plumbing is ancient, and the kitchen and back porch appear to be just barely hanging on by fingertips to the rest of the house. As far as heat, it has none. I mean nothing. It does have four fireplaces with connections for wood heat, so I reckon this winter we'll start that experiment.

At the cabin, I've spent the last few days laying the foundation columns for the wings — the kitchen and bathroom sides. Charlie has been helping me frame all that in. And meanwhile I'm getting my first serious lesson in carpentry. The first thing I learned is: get a good hammer. My lightweight, rinkydink hammer made Charlie laugh. And I broke an old wooden-handled hammer I borrowed from Bill.

I got what both he and Bill recommended: a big rubber-handled, steel-shanked Stanley Rocket. Now this is a hammer. When I hold it in my hand, I feel its authority. So do the nails.

Wednesday Aug. 14, 1974 — This moving business has been very time-consuming and frustrating. It's a wonder we survived the last few weeks. The move was simply terrible: two long weekends with a U-Haul and friend Tony Napoli helping with the heavy stuff. I hated being away from the cabin, but one thing I'm learning this summer — you do what you can do, and be thankful for that.

Friday, Aug. 16, 1974 — And thankful is what I am tonight. Charlie and I worked like wildmen the last three days, pushing through the framing of the north wing, and today was glorious.

Charlie keeps exclaiming "how free this building is..." since we have no blueprints or foreman to answer to. Charlie grins and shakes his head in delight, saying to himself, "I love this. I just love this."

You want a room here? We wave into space. We put a room here. Creating something from raw nothings. Three rooms became wooden reality, emerging from napkin drawings, our imaginations, and the demands of the building itself.

We two worked through sun and rain until that section was entirely raftered. This was no easy task because I decided I wanted the wing extended to make 12-foot-wide rooms, extending the cabin wings by four feet on either side. This meant extending the rafters from 13'8" to 15 feet.

I came up with a method for extending the rafters, and it involved making new lumber work with old round pine pole rafters. Charlie agreed it was OK structurally.

We ran into plenty of problems along the way. For instance we totally forgot the 22-foot-long sill on the north wing. It had to be manufactured as most of that stall on the original barn was completely rotted out. Luckily Billy and Jeanette had donated a bunch of great logs, 5x7s and 4x5s, from an old outbuilding they'd salvaged up at their place.

The 22-foot-long outer beam (with a 32-inch overhang over Selena's bedroom window) was pieced together from two of those beams; big brawny things they were too. We did what's called a lap joint, and pegged it together.

This took longer than expected, but if we learned anything in August it was that things done right take longer.

By the time we got the last of the 13 rafters up, we were accompanied by a light August shower. We didn't care; our energy level was high and we were so nasty and dirty that the rain came as a welcome refreshment.

Another thing: why run in from the rain when you have nowhere to run? And the irony wasn't lost on us — that while we were building shelter from the elements, those very elements were reminding us how fickle and domineering they could be.

We were hot and sweaty when those rains came, and so we just worked on through the glorious shower. The thirteenth hand-notched rafter slid into place, and the sun broke out over us to bathe Sourwood Holler in August tinsel. We cheered and then laughed at each other for looking like two happy drowned cats, hair all matted to our heads, clothes dark with the wood grime of a 150-year-old barn.

We took silly pictures of each other. Charlie looked beat. But I struck an Al-Jolson-in-gym-shorts pose. Ta-da! This place is coming together.

Yes! The north wing rafters are up, rain or no rain. Ta-da.

Roofing 41

Sunday, Aug. 21, 1974 — Time is sweeping implacably through the year's rushing arc; a faceless juggernaut bent on nothing but its own motion. Time, that Fagin, hurls me ready-or-not into fall. Summer now lives in boxes of black-and-white photographs and a red journal that lamely excuses itself.

Autumn creeps lightly into Sourwood Holler. From the cove's green-plumbed depths the cabin rises. Slowly it has sprouted wings. The wings of a great wooden bird spread about her like a mother hen shielding biddies.

We, the Biddy family, work beneath those wings and rejoice in flights of fancy: there'll be a big glass door here, and a skylight here, and....

A week flits by. 200 trips to Henson Timber Co. Another couple dozen supply runs to Hills Hardware. Got any eight-inch lag bolts? And a resulting treasure hunt in the store's basement with Robert Hensley.

The rain tries its best to wash the road away. Silver Creek gargles her incessant watery yodel.

The sounds of approaching autumn wheeze in reedy sibilance. The woodwind section is chaired first by those soprano pipers who know but one note: listen. In the country you can hear autumn coming.

The first turncoat is the sourwood, whose finger leaves play treason to the summer, abandoning her greenery for an August burgundy. Deep scarlet splashes stain the still verdant cove like spilled wine.

The cabin, like a live thing, settles and adjusts to its new setting and the time of year. Already the logs, green with last week's rain, have found their gray exterior again. The cabin snuggles to its bed as summer dwindles. With time, it appears more mushroom-like, and less like a creation of man.

Each day as I leave the Holler, I pause part-way up the hill to turn and look back on the cabin...and sometimes I can scarcely believe that I had anything at all to do with its being here.

Sunday, Aug. 25, 1974 — Now, a week later and August is all but gone. It's been an amazing month of comings and goings, of moving and settling, and of discoveries and frank confrontations with reality. Work at the paper will get intense again this fall and I won't have as much time as I've had this summer.

We've come brickwall up against the stark realization that our log cabin cannot possibly be ready for habitation by this fall. So now work progresses...but we feel a new sort of freedom since the deadline effect has been removed.

42 Hogwild

For the last three days Charlie and I have been involved simply on the kitchen side. Framing in has required much ingenuity, for that side stall had all its logs intact and melding old log beams with new lumber is taking a lot of time and unconventional carpenterial thinking.

The challenge is to take the original hand-hewn logs of the old cabin and mold it with the shiny-bright, perfectly planed and kiln-dried 2x8s from the lumber yard. Where the new regular surface meets the most irregular and undulating old beams — ah, here, Charlie and I are in new water, like ancient mariners with no maps except the warning: "Beyond Here, There Be Dragons!" So, we go slow, and with respect.

Notes to myself: Get 90 feet of 12-inch-wide aluminum flashing for ridgetop, exchange 15-pound felt paper for 30-pound weight, order cedar shingles for 1000 square feet of roof, get 50 pounds of roofing nails.

The wild roofer strikes a heroic pose

Aug. 31, 1974 — All this week Charlie and I have been sheathing the roof with plywood. And I have learned a great lesson in physical science; how to confront a 4'x8' sheet of plywood, grab it squarely with both hands, heft it (and these buggers aren't that light), then balance it somehow, walk up a ladder, and get it onto the roof rafters.

The actual sheeting of the roof has gone quickly, and Charlie has done much of that work. Ready now for shingling. I know it sounds impractical to take the time to roof this big place with individual cedar shakes, but to our way of seeing it, this is absolutely the only type roof to have over an old log cabin.

Sunday, Sept. 1, 1974 — This is glorious fun — quite easy, too, once you get the knack. We started at the kitchen downside, doubling the first course of sweet-smelling cedar shingles to provide a drip shield. Then we established a lovely alternating pattern with the shingles that Charlie said made it look like "you've hired a drunk to help you do your roof."

Wednesday, Sept. 11, 1974 — We have spent five sessions roofing now, and twice Bill has helped considerably. Charlie has been just incredible; he's taught me so much about carpentry and about being laid-back and still getting it done.

And that's what we did today. We actually finished the blasted roof, finished it totally and forever. It was extremely difficult working on the steeper profile of the roof, the ankles took a beating -- not to mention having dreams at night of falling off.

But we persevered — and in the final minutes of today's work, like horses running for the stables, we galloped through the final bits of touching up and finally crowning the roof with a tiptop row of coursing along the ridgeline. My way of celebrating was to take pictures of Charlie straddling the ridgeline like some circus elephant rider.

Bill and Charlie start in on the cedar shake shingling.

Then the last nail...and Charlie climbed down and just sat there in the driveway gazing sort of amazedly back at our creation. The look on his face mirrored the way I felt too. Damn if we didn't do it: all 5,550 shingles, 52 rafters, about 65 logs including two 40-feet-long wing-logs... So we took pictures. Five-year-old Selena shot one of me slouched happily in front of the cabin, exhausted but happy, and waiting to see if she was going to drop the Nikon F or not. She didn't. And if she had, I probably wouldn't have cared. Just too happy to hurt.

Wednesday, Sept. 18, 1974 — Taking a break from the Summer of '74. Charlie says we OD'd on shingles. Maybe so. One thing I know is that this project is an intensely personal journey.

"It'll rain in on you," preached one reader when he saw a cabin picture in the paper. "Looks like you've got the hardest work yet to do," said another. "When you gonna get that place finished?" I'm asked.

I believe I know the answer now: longer than just 90 days, longer than I had ever imagined possible. So long in fact, that had I known, I don't know if I would have taken this project on.

Now, I can't say when it will be "completed." That word smacks of deadlines, and Lord knows in the newspaper business I have enough of those already.

So after a couple of weeks off, Charlie and I will start in at a pleasurable pace with the floor joisting. I'm not going to subject myself to the tyranny of time.

Right now, before I can look ahead too fast, I have to look back at the Summer of '74 when a barn came down, and a log cabin went up. Pioneer zeal or not, I couldn't have done it without my friends. And now, I feel deep down that we're entering a new phase. A time to get down to basics; down to flooring.

Nailing down the ridgeline, Charlie rides the crest like an elephant trainer.

5

Joisting Along

> *"In wildness is the preservation of the world."*
> —*Thoreau*

Wednesday, Oct. 2, 1974 — We watch the cedar shake roof already turn from red to gold to bleached pale yellow as time and the elements do their magic. I'd like to think I'm aging as gracefully.

After a three-week layoff, the Sweatmore Construction Company is marshaling its forces for the Fall Offensive. The cabin awaits its innards. Floor joisting and subflooring come first. Faithful sidekick Charlie has been advanced $100, and we'll go at it this week.

My body has recovered from the summer's wounds. The gash caused by that nail during the barn-razing has left a curious scar on the side of my torso: it looks like a tiny figure who appears to be in the middle of buck dance step. I wear it like a tattoo, a dueling scar.

My shoulders have stopped aching, and hammer-banged thumbs are all healed. That probably means it's time to go out and bang them again.

Thursday, Oct. 3, 1974 — New wine! Walking up above the cabin this morning to get a good view back — I noticed on the ground a round, purple-black marble-like thing...there — another.

My gosh, I realized — scuppernongs — everywhere!

The ground was littered with the fallen deeply-ripe grapes, as if a kid had spilled his bag of purple marbles.

Selecting a dark black beauty that was fairly bursting with expectation, I chomped down on autumnal serendipity. What a fruit. It spread through my senses like — yes — new wine.

An incomparable taste. No wonder as a child I never cared for this variety; it's an adult taste of fermentation. Each bite is a heady drink of nature's spirits.

Looking up, I saw above me in the spindly pines, a grape vine

46   *Hogwild*

loaded with purple beauties. And they were falling like manna from heaven (sweet purple hailstones) and bouncing on the pinestraw floor like so many cheery unbreakable Humpty-Dumpties.

I got down on all fours and gladly scrounged all I could, recalling what an anthropology professor once said — that the lowest economic form was the "pickers and gatherers." Ah well, I've come to that, have I? The woodbutcher knows he's getting somewhere when he uses a salvaged 40-penny nail for a pipe-tamper — and starts writing on wood.

Having no notepaper out here today, I was at a loss — until I divined that cedar shake shingles might make a fine tablet. Et voila! I'm writing all this with ballpoint pen on leftovers from the roof.

Now I'm sitting in the "kitchen" and I use that word loosely. Right now, it consists of the bare log outline. I'm perched on Big Mama, the 10x10, 22-foot-long beam that spans the length of this room-to-be. She's a crusty gal, a hand-

Taking notes on a backwoods tablet.

hewn beauty who could have given the Clairol ad men their inspiration about getting older and better.

Sourwood Holler has lived up to her name. The sourwoods are rampant with diamond-hung pendant blossoms. And the leaves are deep Sangria against the royal blue sky of morning.

A new sound: that of gentle descent. Leaves clippety-clopping from their perches. A flock of tulip poplar birds are engaged in free-fall. Every leaf that falls lets in a little more sky, making things more airy and vastly blue.

Thursday, Oct. 17, 1974 — Charlie hasn't shown up. I'm beginning to think it'll be a while before I see him. The joist beams have been delivered and I reckon I'll have to do it by myself.

Down by the creek the blue and white striped tent is still standing. Today it reminds me of a knight's campaign tent. Well, that fits. I joust. I joist. Brace yourself, Sir Cabin.

Saturday. Oct. 19, 1974 — I have started painting the town house in Rutherfordton. We settled on a deep Wedgwood blue to cover the off-dingy white that it was. And this color, plus the place's poor construction, has led to a name for the old house: the Blue Shed.

Wednesday, Nov. 6, 1974 — The campfire smell permeates the sunny Sunday afternoon where by the Silver Singing Creek I am pecking tentatively at the old black Royal I've lugged down to the picnic table.

A fire started just for company in the rock fireplace returns a Sunday *New York Times* fittingly back to the elements in sunshafts of sweet smoke.

The campfire cackles in post-Halloween merriment, sending witchy sheets of heat to distort the far hillside as if viewed through old window panes. The fire speaks of cookout breakfasts and late-night sing-arounds. The creek mutters of wading and boyhood dam-building.

Now here's a fine contradiction: fire and water combine for a duet of opposites.

From the lower side, the cabin looks like a giant mama hen trying to get airborne.

48   *Hogwild*

Overhead, an Indian Summer breeze, like some celestial whisk broom, is sweeping away the leaves of Sourwood Holler leaving gaunt gray guardians (Rodin's "Burghers of Calais") standing about as if caught in a frieze-frame of my camera shutter's one-thousandth-of-a-second blink.

Leaves like balsa gliders pour out of the azure. I lie on my back and play little games. Will one land on me? Some spiral down like helicopters, others take purposeful, slanting dives to their landing strips.

On this impossibly beautiful afternoon all I'm wearing is tattered jean cut-offs (the ones which Maggie lovingly embroidered "Zonkie" on one rump). I have to marvel at the temperature; you can't tell where your skin stops and the air begins.

From the battered, long-handled tin dipper hanging on the little hickory tree limb, I take long draught of spring creekwater. The taste of it is like a sweet, cold kiss.

The two good dogs flounder happily about downstream, and now Ruby comes to thrust her sopping wet, graying muzzle to my bare thigh; I return her thoughtful, devoted gaze. By god, it is surely good to be alive this November day.

Sunday, Nov. 17, 1974 — I am three-quarters of the way through joisting the kitchen now — and because I'm working with short spans of five feet and less, I have discovered it is something I can do without help. I guess Charlie will make it up to me someday.

I put 2x8s on 16-inch centers. Getting those rascals level has taught me a lot about notching and fine work.

Today while working I happened to look down at the ground — something had caught my eye in the dirt beside the foundation. I reached down and brushed away the soil from the small white object.

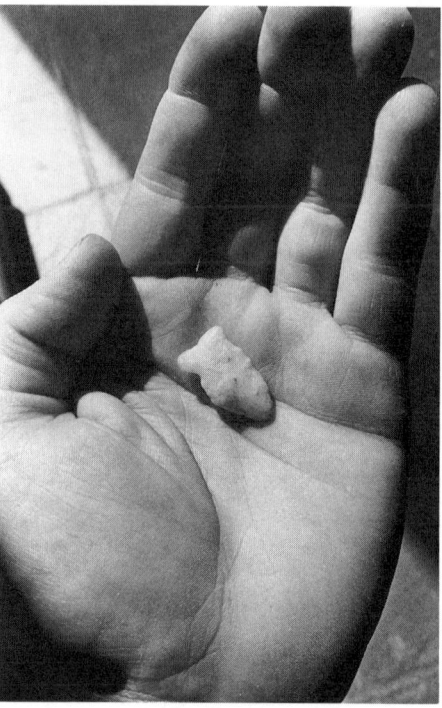

A good omen from the past.

An arrowhead. I was stunned. I stared at the perfectly carved white quartz relic in my hand as if it was a message or a gift from one ancient authentic native craftsman to this would-be woodsman.

To think, not only that my cove was inhabited by Indians, but that right here, exactly where we chose to build a log house, an arrow fell, and lay here all those years — as if it was here for me to find. It's enough to make me believe in signs. A good omen, this.

Sunday, Nov. 24, 1974 — Maggie, who is working as a roving folksinger with the county school system, has discovered a rockmason for me. Kurt Kaltreider, a scholarly philosopher, mid-twenties I'd guess, is teaching up at Union Mills Elementary School and rocking in his spare time. He's agreed to do the fireplace and chimney.

Last week he came out and helped me with the kitchen joisting, and we've become buddies. He refuses to take any money. Says this is as good for him as it is for me. Now that is a friend. And right now I could use one.

Next on our agenda is to order 13 tons of Carolina gray granite from an old rockman he knows way up in Marshall; that's higher up the mountains of Madison County.

Christmas Day, 1974 — A year of a family odyssey, of time and moving, of old houses and log cabins, a summer we thought would never end -- but then it ended before we were ready.

Looking back now at the events we lived through during that one trip around the sun, how it all stands in clarity, insulated like Vonnegut's "bug in amber." How'd we get it all done? Why didn't I get more done?

The other night I made a woodblock print of the cabin. It's very primitive, but somehow satisfying, and appropriate because it's a woodcut. Putting dreams of wood on wood, ink to wood, and wood to paper. Circles within circles.

I hold the cabin woodcut in my hand like a doll of a dream girl. The cabin is out there, with its pale gold roof rising from gray Sourwood Holler. I'm there, but I'm not there. The place anchors me, frees me as it tethers me, stirs me as it becalms me.

Lao Tsu reminds me to hold holy all things natural and mark this Christmas quietly, to find a measure of Quaker peace within:

> *The highest good is like water:*
> *It gives life to the ten thousand things,*

A woodcutter's log cabin

> *and does not strive;*
> *It flows in the places that men reject and so*
> *is of eminent virtue...*
> *Hold fast to the center*
> *for that is the only true way*
> *of being whole....*

Saturday, Dec. 28, 1974 -- We have been heating with wood at the Blue Shed this winter. In the living room-kitchen, we're using a Franklin stove, which is attractive but totally inefficient. For Selena's room Bill helped us find a Fisher "Baby Bear" stove. It's the most air-tight thing I've ever seen. For wood, I'm toting scraps back from the cabin and picking up packing crates around town. I've got a hunch I'm not doing it as well as I could.

This week I finished joisting the kitchen, and tomorrow I'll start sheathing it with 5/8 inch plywood. I'm really looking forward to getting on with things.

I've been re-inspired by books Maggie gave me for Christmas: Eric Sloane's *An Age of Barns,* and an owner-built home book full of photos and philosophy, *Woodstock Handmade Houses.*

A couple of quotes from the latter:

"The more builders, the less bombers."

And: "...the important thing is how these Old Places were salvaged from abandonment and love — or lack of... that's real ecology — revering materials and the long-ago labor of these self builders. Really look — and you can see what those old-timers wrote with an adze on the side of a beam."

And as if the authors were reading my mind: "Every shelter was once just a dream."

Dreaming is one thing: now let's get off our duffs. To do:
- subfloor the kitchen.
- block and joist rest of cabin.
- get remaining timbers and beams down from barnsite. Stack and cover with old tin.
- build good bridge across Silver Creek.
- get rocks
- begin rockwork
- get blocks, cement, and make base for firebox/chimney
- measure spaces for doors, windows, and fireplace
- cut out holes in logs for same.

Rockmason friend Kurt Kaltreider helps joist on the kitchen.

6

Wherein Our Hero is Floored and Builds a Playhouse

"One truly understands only what one can create."
— *Giambattista Vico*

*"Thirty spokes share the wheel's hub
It is the center hole that makes it useful.
Shape clay into a vessel;
It is the space within that makes it useful.
Cut doors and windows for a room;
It is the holes which make it useful.
Therefore profit comes from what is there;
Usefulness from what is not there."*

— *Lao Tsu*

Saturday, Jan. 11, 1975 — A new year, a new page, and a new phase. Today I finished subflooring the kitchen. Tomorrow I do the block supports in the center section for the 18-foot-long joists.

Wednesday, Jan. 22, 1975 — The rocks for the chimney came today. Both the load and the guy who delivered it were something else. Tom Caldwell ("Old Tom," Kurt calls him) was as bent as a question mark. As soon as the grizzled mountain man from way up in Madison County got out of the flatbed truck, he eyed us up and down, grinned toothlessly and from his spacious wool jacket pulled a bottle of Wild Turkey.

"If'n you boys are gonna help unload these rocks, you're gonna need this," he croaked, proffering the bottle of liquor.

Whew-EEE! If that stuff didn't make me want do a "double back-flip and run a half mile..." We sort of got loaded unloading.

The rock itself is the loveliest rock I've seen — Carolina gray granite,

Wherein Our Hero is Floored 53

Thirteen tons of Carolina gray granite arrive from Madison County via "Old Tom" Caldwell, right.

After the Wild Turkey it's time for a smoke.

Tom called it. And now I've got 13 tons of the stuff strewn around the dogwood tree beside where the chimney will rise.

Thursday, Jan. 23, 1975 — A reverence for things natural. One of the things I've learned along the way is that I've always had a profound respect for rock and wood in its natural and aging state; man cannot improve on nature.

Bill tells me the Japanese have a word for this: "Wabi," roughly translated as "the artful state of decay." I have come to regard the barn itself — just as it is — as some sort of a temple.

This is good and bad; while it means I respect the structure and its innate identity and karma, it is also true that we can't live in a barn. So philosophically today's work represented a giant step forward. It was a little traumatic cutting out the 6x6-foot hole for the fireplace and chimney with my Stihl chainsaw.

The first cut is made and the firebox is in position.

It's the first time I've actually cut into the barn, altering the log structure in any way, and there was something frightening (would it weaken the cabin? Will it fall down?) and almost irreverent about it.

But my fears were groundless. The operation went fine — and besides, it had to be done. Sort of like having a tooth pulled to make way for something new.

Today also I built a little bridge across the creek, lugged remaining logs from the barn down to the cabin, and covered them with tin.

Saturday, Jan. 25, 1975 — I got a good start on joisting the north wing, which includes Selena's room and the bathroom. Don't forget: order two tons of "Columbia white" sand from Robbins Brick Company, plus 10 bags of No. 1 Portland, something in which to

Wabi -- the Japanese have a word for "the artful state of decay."

One horsepower: Bill uses Tony to haul some more logs.

mix the stuff called a mortar box, and old shovel and hoe I don't mind ruining since I'll be mixing it by hand.

Also, the "Heatilator" firebox has been delivered; it's one gollywhomping hunk of steel.

Friday, Feb. 7, 1975 — I'd call this appropriate energy use: Bill Byers wanted to see how well Tony the horse would do as a draft animal, so he hitched him up and dragged a couple of donated logs down to the cabin. No tractor's rumble, no noxious fumes. And it was probably the same method used around 1850 when the barn was originally built.

Just good honest one horsepower. We call it low-tech.

Saturday, Feb. 8, 1975 — Kurt and I started in on the blockwork for the firebox today — Kurt laying block and me mixing and slopping out the mortar. Something as massive as a rock chimney starts humbly enough with earth moving: starting with sand in a borrowed wheelbarrow and digging out earth for the footings.

There is something so kinetic about this building process. I stand back from the cabin, hold up my hands to cup an imaginary monument of rock, framing its image with the hands that will build it...right...there.

56   *Hogwild*

Click.

The mind's shutter winks and the image is briefly but indelibly imprinted on the emulsion of my inner eye.

Saturday, Feb. 15, 1975 — Hair? what has hair got to do with all this? It seems the more the cabin goes up, the more hair falls out. With that, plus the weight lost working on the cabin last summer, there's no telling what I'm going to look like when the house is finished.

I can see it now: I started out fat and fuzzy, and when I'm done here someday, I'll be bald and emaciated — having given my all for a legacy in wood, rock, and mortar.

Kurt starts with the firebox base.

Today, while Kurt laid block I placed the first of the 16-foot, 10-inch-long roughcut 2x10 joist beams for the main room. We had to work together to determine where the floor level would be in relation to the fireplace, which we decided

In the living room, I dream after putting up the first joist.

would be raised about a foot off the floor. Happily, the two bottom sill logs (X-1 and Wild One) are square and larger than their buddies that make up the wall. This allows me to lay part of the joist notched on the log, which adds greatly to the floor's stability.

Maybe this is Carpentry 101 and all builders already know this stuff...but I don't have any knowledge or experience in this field. Every challenge I come up against I just have to think through and then wade through. I go a lot on faith; I'm like a dowser looking for water. I have to trust the wood to take me where it wants to go.

So I spend a lot of time just standing around, gazing into space...thinking, thinking. But what I'm doing is imaging. It's like a giant chess game I'm playing with an entire house; each thing I do is like a move on the board, and I'm projecting into the next move: how will what I do today affect what I do tomorrow?

Sounds like a model for responsible living; little wonder I feel myself growing so.

Sunday, Feb. 16, 1975 — Step Number One for Building Log Cabin: get a tetanus shot.

The ersatz pioneer rediscovered that rule again this week. Earlier, Kurt had slashed his finger on a rusty nail and went and got a shot. But it wasn't long until he was putting the serum to good use again.

Yesterday I had propped several large upright "Zonk" logs in place between the kitchen floor and Big Daddy, the main overhead beam running the length of the kitchen. Kurt was rummaging around in the kitchen, not realizing that the Zonk logs were just free-standing.... He was kneeling on the floor going through some box when he bumped the bottom of Zonk 3, a 10x10 hunk of solid heart pine — now crashing down soundlessly on him.

Kurt must have sensed doom, for he glanced up just in time to spot the 100-pound monster whispering down right at his head. He vaulted backwards and the log missed his head — but caught him amidships and pinned him to the kitchen floor, where he lay gasping for breath while I came running.

Once he had climbed out from beneath this would-be sword of Damocles, more dazed that injured, Kurt demanded pointedly, "Why'd you name that log Zonk? 'Cause that's just what it did to me."

Label a log "Zonk" and that's what you'll get: zonked. I guess I'll have to be more careful with my appellations around here. Things have a way of living up to their names.

58   *Hogwild*

Saturday, Feb. 22, 1975 — Spring makes timid entry into the Holler with spring peepers singing tentative songs ("wheep-wheep!") down in the marshy end of the Soccer Field.

We finished the block support for the firebox, and it took four guys to heft the piece of steel into place. Then we started on the hearth base inside the cabin, and we're filling it with junk rock and freebie brick I'm picking up anywhere I can find it.

We're ready to start rocking soon. I can't wait.

Saturday, March 1, 1975 — A remarkable day during which I finished the joisting of the creekside end of the main room. I guess I ought to be calling it the living room but it's so big — it's the 18' wide, 22' long core of the old barn.

Today included a couple of miracles (which I have learned almost to expect now). Earlier, I had been concerned how the deck could be affixed to the house. But today I came up with the idea of starting the deck at the northeast corner of the cabin — and resting part of the deck joists on the foundation. These deck beams, which run underneath and perpendicular to the room joists, by their sheer weight alone support the final two main room joists. I mortised them all into place.

The result is a cantilevered effect, I believe it's called. The weight of the two main room joists on top force the four 12-foot-long deck joists onto the fulcrum of the cabin foundation. It's about the zoomiest thing I've come up with so far.

Thursday, March 6, 1975 — Transactional analysis guru Sam Keen would have liked today. Sam says: don't deny a day or a circumstance the possibility for novelty.

Kurt lays the cornerstone of the chimney.

Wherein Our Hero is Floored 59

When Kurt came down to the Holler today, I figured we'd be doing more inside blocking on the hearth — but instead he said: "What the heck! I want to start rocking."

So just like that — we did.

The laying of the first rock was maybe not as big a deal as an official cornerstone laying. But still we marked it with picture-taking and cheers.

Since I had ordered two and a half tons of good sand ($29) with the new mortar box, mixing mortar is much easier now than it was with the Byers' borrowed wheelbarrow. Kurt has taught me how to mix mortar for rock: here's his recipe — take half of a 94-pound bag of Portland plus 15 shovelfuls of good sand, mix to even consistency, then slowly add water and stir to the proper feel. For rockwork, Kurt likes it kind of dry — especially compared to "wet mud," as they say, used for blockwork.

I learn also that mortar for rock "cures" over a period of 24 to 48 hours, and also that the most important thing in rockwork is not the mortar (as the novice supposed) but the fit — and especially gravity. All the mortar in the world won't hold a sloppy fit in place, Kurt tells me.

By the end of the day, Kurt had laid two courses all around the base of the chimney. "Face-laid," it's called, because the broad "face" of the rock faces out, as opposed to "flat-laid." We chose this style quite naturally, because this rock is so beautiful that it would be a shame to hide any of it. But this kind of work requires infinitely more patience and skill — even I can see this — because a whole day's work can simply collapse if you bump one rock before it is seated.

Between the rock and the firebox, we're putting in about 12 inches of junk rock to fill

It's slow going, but the "face-laid" rockwork is lovely.

Down at the Big House, Artist-in-Residence Deborah Compton gives a puppet show for the kids of all ages.

and insulate. This is going to be one bohunker of a chimney.

Friday, March 7, 1975 — Well, how about this: the newspaper is growing so nicely that we're contemplating buying a larger office uptown around the corner on Main Street. And the place we're looking at is so big that we can put a press (we've got to buy one first, dreamer) in the basement, and get to design our own offices and darkroom as well. After six years of working crazy hours, I reckon we've earned it.

For my office I think I'll use all that gray stately oak siding that was originally on the old log barn's stalls. I bet when it's dried out and wire-brushed, it'll be a fitting backdrop for black-and-white mounted enlargements.

Saturday, March 15, 1974 — After finishing what's called "bracing and bridging" between the joists (X-shaped connectors made out of 2x4s) I made great progress in sheathing the main room with plywood subflooring.

My whole perspective changes now that I have a floor to walk on. Remember, all I've ever been used to is a dirt floor in this log barn. By golly, having solid wood to walk on — and it so sturdy underfoot — is downright thrilling to a kid who a year ago practically didn't know which end of a hammer to hold. Yes!

"Yes." Why did I say that? The word has become a motto around our house to symbolize "Yes I can, yes I can."

I am The Little Engine Who Can. Maggie has taken up the hue and cry, and somehow found me a folder that just says "YES" on the front. Into this I put napkin drawings, legal pads full of designs, catalogs full of ideas and bills from building supply places.

On top of that, she's taken my beleaguered and fading red gym shorts that I wear out here every day (and wash every night) and embroidered a small "Y-E-S" on the left haunch.

When I'm out here totally alone, I have my Yes Book and my Yes Pants to silently urge me on. And there are times when I need all the positivism I can get.

Saturday, March 22, 1974 — We made it through our first winter heating with wood — and I for one appreciate spring's coming this year in a way I never have before.

We Hogwilders gathered last night at Bill and Jeanette's to observe the Vernal Equinox's arrival. At our own Rites of Spring, we woodchuckers chuckled and clapped each other on the backs for having chucked — oh, say three cords apiece through the long winter.

On the new living room floor we spread our cots and spent the night in the cabin for the first time — no windows, no chinking... it was like camping in one big log tent.

And this morning we were awoken by birds going berserk with song: doves hollering for you to plant your corn and mockingbirds just singing fit to bust over April's coming. There are some kind of teensy warblers whose high-frequency twee-to-weet is so far up the octave that their song practically hurts the ear... and a pair of industrious flycatchers is building a nest under the eaves of the cabin's overhang.

My Quaker granddaddies call to me from the past to Be Still and Know. So I listen a lot out here. Silver Creek joins with those small-fry frogs in a welcoming anthem to the return of Life to the Land.

I learned from reading Foxfire that the way to build a cabin is to erect the complete structure first...and only then do you cut your doors and windows.

I had done some careful measuring, based on using the salvaged doorframes and doors out of the tenant house we've been using. This morning, Maggie's dad, Ralph, came by — and since he's an old mountain boy with a sure hand with the chain saw, we let him loose on the project.

In little under 30 minutes, he had done the job — with me directing and grabbing each cut-out log as it fell, propping and bracing the standing walls with chocks of wood and us shouting over the chainsaw's razzmatazz. Maggie, with her hands in her overalls, watched in stunned amazement as Dad cut the two main doors, one leading out into the kitchen/entrance and the other to Selena's room and the bathroom.

When it was done, we could walk from room to room for the first time, freed from the horizontal "bars" that had made the central core of the cabin something of a prison. Now we were given a feeling of continuity...so we walked around the place for a while, just getting the feel of the newness and free space...all the napkin drawings in the world can't compare with the actual sensation of walking through your dreams like that.

Oh, we'd say, so that's what it feels like to walk from here to there.

Cutting the doors: Maggie's dad, Ralph Palmer, takes chainsaw to 150-year-old logs. Suddenly, we can walk through walls.

Wherein Our Hero is Floored 63

Then, when this was done, I made the mistake of saying something about how nice it would be to build Selena that playhouse she's been moaning about for her sixth birthday, April 11.

Dad, who was feeling his oats from the project just completed, said, "Hello, let's do it." And so the three of us pitched in, clearing a spot across the creek that faces the Soccer Field, laying a quick block foundation and putting together an 8'x12' base frame, flooring it with plywood. Whambang, nothing to it, thanks to Dad's energy.

We can now walk from room to room for the first time. We have a new sense of space.

Sunday, March 23, 1975 — It occurs to me today that what I'm doing is building far more than just a playhouse. I should have guessed it earlier. This place is just as important to Selena as the cabin is to us.

And to Maggie, the playhouse is a reincarnation of her childhood place, called "the Bluehouse," that sat up in the apple orchard between her parents' and grandparents' house. That was really the genesis for this project: Selena hearing Maggie talk wistfully about her beloved Bluehouse — a hideout for little girls and dollies and make-believe beside the enchanted forest.

So the Playhouse takes on added significance, and we race

The basic frame of the Playhouse is up, as Maggie feigns total exhaustion.

for time against the birthday deadline. Today we threw up a basic frame, and Bill happened by to teach me how to build rafter sections. After he left and I was scrambling around bare-backed in the fine spring day, artist friend Bill McCullough (who lives over the next ridge just outside Hogwild) stopped by and wanted to help. So we had ourselves a Playhouse-raising. By the end of the day we had that sucker up and the roof sheathed and the shingling started. It's amazing how fast a little place can go up.

I was talking to Bill Byers about cutting out the doors at the cabin, and upon counting the rings, we found the logs to be 100 years old when the trees were felled.

From our research we've determined that the barn was well over 100 years old when we found it. Subtract another 100 years for the age of the logs when cut, and that gets you back into the 1700s, prompting Bill to muse, "That gives you a graphic idea of how really young this country is."

That means the logs of this cabin were young pine trees in the year 1776 when my ancestors were living in central and eastern North Carolina, and — who knows? — packing powder in muzzle-loaders to take on King George III.

Talk about Bicentennial vibes...I can see for myself the broadax signatures of those old-timers every time I look at one of these old logs. And I say to the logs: you guys were growing right here when the Indians hunted around your bases, camped in your shade, when the mountain men whipped up on Ferguson at Kings Mountain... and I am Vonnegut's Billy Pilgrim, "unstuck in time," staring at a wall that is anything but blank

This is what Eric Sloane is talking about when he speaks of "a reverence for wood." Amen.

Saturday, March 29, 1975 — The Playhouse is coming together quickly. We've added windows bought at a salvage place in Tryon, siding from Dr. Hyde's treasure trove from Jonas Ridge, inside paneling Maggie got down in Shelby, porch uprights and banisters salvaged a couple of years ago from a house being torn down in Forest City on Cherry Mountain Street.

Friday, April 11, 1975 — It's been a frantic week trying to get finished in time for Selena's party tomorrow. It's taken a group effort, with

Now the siding's up, and there's the old hen in the background.

Selena helping along with new friend Deborah Compton (a children's theater artist-in-residence with the public school system). Jeanette donated blue curtains, and we've painted the Playhouse smoke blue in honor of the original Bluehouse, adding red shutters and white columns and banisters for a real All-American Bicentennial Playhouse.

Saturday, April 12, 1975 — Somehow we finished in time for today's birthday party at the Playhouse, and it was grand. Friend Andy Morgan, another theater artist-in-residence, gave Selena a Sad Little Clown face with his theater make-up, which is what she wanted from him. Seven other little kids from the neighborhood came out and had a fun day in the Holler.

Number One Daughter reminds me to add that she has named the big hill that forms the opposite side of our cove and looms above the Playhouse. Selena named it "Slickey Mountain," after she and Hannah tried walking up it but kept sliding back down. The two little girls, carrying their dolls, inform me seriously that they believe Slickey Mountain to be home of "real fairies."

And to that I would have to add, naiads and dryads, too.

On the day of the party, kids (and Deborah and Andy) pose for group portrait. Selena (far left) opts to come as a sad-faced clown — a face of Andy's making.

Sunday, April 13, 1975 — A day for reflection. Spring has come romping across Sourwood Holler — all except the place's namesake has burst in splendor in the deep silence of Hogwild: bloodroot first, then pinxster, flame azalea, phlox, trillium, aster, daffodils, shadbush (called "sarvice" by mountain folk) — plus our cove is awash with legions of irrepressible dogwoods.

In an alcove of old log rafters above the kitchen a family of crested flycatchers is nesting among the beams. So somebody has beat us to the punch of living here. Well, I'll respect squatters' rights.

Nothing has happened recently on the chimney. Kurt got about three rounds up, about knee-high, and then got involved with opening a health food store in Rutherfordton, so I don't know....

One crisis at a time, please. This week our newspaper bought that place uptown and we began moving. Reckon Hogwild will have to take a backseat for a while.

Friday, April 18, 1975 — I feel like I'm in italics. Slanting visibly forward. This week I actually got to help build my new office. Each one of the three co-publishers designed his own office, and naturally the old cabinman had to have the rustic look. So I paneled two walls with the old oak siding off the log barn and did wainscoting along the third wall with thinner old barnwood.

This ancient stuff had been rescued by Maggie after it lay in the honeysuckle for a while. Maggie sorted it, de-nailed and covered it. I hauled a load to the Blue Shed and steel-brushed it and let it dry out before carrying it to the paper office.

I think it looks great — along with modern track lights, a fourth wall of bookshelves and built-in layout desk, and plush burgundy carpet. But I have caught a lot of ribbing from friends and workfolk passing through.

Our electrician, Al Buchanan, has taken special delight in teasing me. While installing the track lights in my office, Al wondered out loud, "I can't imagine why you want lights like this in a barn." Not long after that, local radio personality Uncle Bud wandered in, and upon entering my office with its deep carpet, he grinned under that pencil-thin mustache of his and said, "feels like walking in cow manure."

Well, what the heck — by the time I'd added my old rolltop desk, wing-back editor's chair and wood cookstove (perfect for filing negatives), the new office did have the feel of a barn away from home. Or, "barn again," so to speak.

Sunday, April 27, 1975 — It's night and we're back at the Blue Shed after two wonderful nights at the land. Early in the evening a full spring moon ballooned softly into the April sky while the Sourwood Family and best friend Danielle Withrow sat around a small woodstove I've hauled into the airy main room (just for atmosphere) and talked late into the night.

We slept on cots with the moonlight streaming in through the cracks of the unchinked cabin's logs, a fine cedar roof overhead, and a sturdy floor beneath, and around us those old logs in a xylem and phloem huddle.

The next morning, after a hearty country campfire breakfast, Maggie and Danielle practiced some mountain buckdancing in the sandy campsite by the creek. Maggie is an old clogger from way back, while Danielle has just learned. Me, I feel good about walking without falling over my gunboats, much less dancing. But Saturday must have been my day. Danielle broke the dance step down into a simple shuffle, stomp-stomp-stomp (change feet) shuffle, stomp-stomp-stomp — and before I knew what I was doing — by cracky, I was clogging.

We just spent our first night at the cabin, sleeping wonderfully on cots while the full moon streamed through the cracks.

We moved up to the cabin where the floor was smooth and perfect for this sort of foolishness, and there, the trio of buckdancers beat a lively tattoo on the snare drum floor. Must have made quite a bit of a ruckus because later that day neighbor Bill said he could hear us dancing all the way down at the Big House.

Saturday night was blissful again; a full moon night thick with whippoorwill songs and spring peepers lullabying. Today when the temperature soared, I laid my cabin tools aside and reverted about 20 years, spending the afternoon in the ageless art of Creekmanship.

That consists of dam-building, creek-dredging, fingernail-mashing, rock-hauling, ankle-wading and back-burning. The perennial nine-year-old loose on the land. The kid who never got his fill of playing in creeks is making up for lost time. One of my definitions of Heaven consists of unlimited mountain creeks and days like today in which I can wander happily making rock dams and causeways for eternity.

Saturday, May 31, 1975 — I count the man lucky who gets to participate in not one, but two log cabin-raisings in a single lifetime.

Bill's younger brothers, Burwell and Stuart, who are both in their early 20s, have been camping out over on their land on the eastern side of Hogwild. They both work for their dad, Jim Byers, at North State Gas Company, and during their delivery rounds on the big Carolina blue propane gas truck, they have been busy scouting for log structures all over the county. Earlier this spring they put up a couple of log barns from 15 miles away over in Sunshine and Golden Valley.

These gentle, easy-smiling laid-back boys embody the spirit of our Hogwild extended family that cares and shares, works and plays together and profoundly loves this land. When someone needs help or is into a major project, we all just drop what we're doing and head across the mountain.

Another cabin-raising. Burwell has a place going up — with the help of Erwin Johnston, Stuart, and Ricky Robbins. That's Burwell on the right.

This morning the word went out: cabin-raising at Burwell's! I walked from my place up to Bill and Jeanette's Big House, following the old roadbed that winds down into Bill and Jeanette's bottoms, across Silver Creek and back onto my land and the road that skirts the little two-acre bottom we've got and have come to call Teardrop Field for its curious shape... there the roadbed closely parallels the Rocky Broad River, which this morning was running wide, deep, and green. Every time I walk thisland I am impressed again with just how big 300 acres is.

The road turns up the hill, away from the river, and goes past an old sawmill, where Burwell's land starts. From across a sloping field you can see the two barns the boys have put up and the rock piers of the new cabin.

Burwell and Stuart were already barebacked and working with friends Erwin Johnston and Rick Robbins on the lower couple of courses of a cabin that they estimated to date back to the 1830s.

We five had our hands full with the job of heaving those huge logs up, up and up again higher. Five guys mustered just barely enough muscle. Some of those logs were eight inches thick and twelve inches deep and thirty feet long. When it's that old heart pine, that's a lot of log.

When we're all gut-busting to manhandle a big log into place, straining to keep our balance and keep our end

Burwell enlarges a door on the cabin while Rick Robbins watches.

up — it's a virile, team experience. Football for keeps. If we can score and push this thing across the goal then we get more than just six points; somebody gets a house to live in. Without Burwell and Stuart last summer, I couldn't have built my place. So it was satisfying to be able

Stuart watches as Burwell uses a go-devil to knock a log into alignment.

Wherein Our Hero is Floored 71

to return the favor today.

It got hot quickly today. Their cabin site sits high on the southern flank of the far side of Laurel Mountain and gets the sun's full blast in the morning. Sweat ran down into our faces, stippled our backs, and made rivulets down our chests in the grime of 150-year-old logs. We were soon filthy but having great grimy fun.

When you build a log cabin it is a total body experience: you use practically every part of your body to get a log in place: not just your hands and arms, back and shoulders. It reminds me more of rock climbing than carpentry. You might use your thighs to balance a log briefly while you grip your perch with your knees like a horseman. You have to pay constant attention to your feet and your toe-hold. And because it's a team experience, you're always monitoring where everyone else is and what they're doing. We became profoundly interdependent; no one person can single-handedly take on the weight of a log. So we learned to move like five men in a sack race. We became a single body.

With the logs all up, we rewarded ourselves with brews and hit the old swimming hole on the Rocky Broad. As John Denver says, "Thank God I'm a country boy!"

Then we hit the Rocky Broad, where the waters refresh and regenerate us.

7

In Which We Chink and Daub, and Win a Press War

"Qui s'excuse, s'accuse." *
— Gabriel Meurier

"There is no heavier burden than a great opportunity."
— anon.

June 20, 1975 — We have retreated to Brigadoon by the Sea; the first week-long vacation I've had since the paper started six years ago. And even this had to be rescheduled around Little League playoffs — such is the life of a small-town newspaperman.

Listen: re/treat... to savor a dessert once more. And also, to beat an orderly retreat from the Western Front. Western North Carolina and the cabin, to be exact.

We are at a little-known fishing village on the Carolina coast with the anachronistic name of Surf City ("two girls for eeh-vree boy...") I doubt there are even two "girls" on this entire beach — at least not the kind of which the Beach Boys crooned.

The ocean is the leading lady here, and none may compete. It has just finished raining and my red-jacketed journal page is splattered with blue flecks as the ink runs before the waning storm. A cooling breeze. The booming thunder out over the Atlantic. I am burnt but relaxed for the first time in ages.

It's as if the tension and anxiety of newspaper and building deadlines, some real and others imagined, are flowing out of me with the receding tide.

The cabin is in a resting phase. We've done little or nothing since finishing the Playhouse in April, May being taken up with moving the

---
He who excuses himself, accuses himself.

newspaper office and a full month's worth of weekend singing at weddings and at Union Grove fiddlers' convention.

I turned 30 this month — and it's as good a time as any to stop on the decade's nexus to take stock of where we're going with the Hogwild dream.

I've always heard it said: it's all a matter of style. I'm beginning to suspect that growing up is largely a matter of learning to deal with compromises, of reconciling dreams with the constraints of reality.

My rockmason, Kurt, has had a boy, Peter, by his fair dancing wife, Sallie, and also opened a health food store in town. So absolutely nothing has been done on the chimney since that flurry in March. And how can I expect more? We have only a gentlemen's agreement, and he can do whatever whenever since he won't accept pay.

Secondly, Maggie and I arrived at the agonizing decision that we must not go in debt for further construction materials — thus setting back the cabin's completion. Therefore we have elected to put off spending $1,500 for Andersen Windowalls, $1,000 for plumbing, and $1,000 for wiring until we can afford it.

After all, why'd we go Hogwild in the first place? Part of it was to be as self-sufficient as possible, and that meant financially too.

I think I'm relieved.

June something, 1975 — I'm not even sure what day it is today. How unlike me not to be grounded on time and calendar.

One of my transactional analysis mentors, the author Sam Keen, preaches the doctrine of the Vibrant Present. Okay, here it is: curtains billowing inward from a sea breeze, towels drying on the porch waving back at the wind, beachbooks stacked in sandy array on top of the refrigerator, a pile of beach memorabilia and oceanic icons cluttering one end of the dining table, a usually busy camera empty now and making a great paperweight....

"We live stripped down at the beach —" Maggie's wisdom for the day. And she's right. A minimal amount of clutter gums up the vacational workings. Guilt is left behind. Nothing in Surf City we can improve on. No floors to sand, no log walls to chink, no chimneys to fret over, no cabins to feel accountable for. The follower of the Tao is reminded: Have not — Want not.

But in truth, we know our wanting is only on vacation too.

So what does a one do on a woodbutcher's holiday?

"Have a ball! And if you don't have a ball — it's nobody's fault but your own!" shouts a fat lady from New Jersey at her brood next door.

She's been issuing have-fun-or-else orders at her cowed clan ever since they collided with the Atlantic yesterday. Nobody talks back, I notice.

Secretly, I take her advice to heart. I have shed my watch and other conventional monsters that mark the tyranny of time, and so doing, I have began seeing the "ten thousand things" that the Tao speaks of. Thousands of little rounded pebbles, polished by eons of oceanic tumbling land pounding. When Maggie suggested that we might use these jewel-like obloids in daffodil planters, I was forcibly struck with the notion of collecting sacks of them.

After two days of picking up hundreds of these rocks, I began wondering what on earth I was doing. Am I that production-oriented? So obsessive-compulsive, that I am compelled to throw myself headfirst into some artistic endeavor — even one as humble as gathering the smallest pebbles on the beach?

Yes, and then...maybe it's something more than that. So simple and perfect are these lovely objects that I am led to a new thought, as if the Atlantic is too vast and too gross an object for a human to truly understand and embrace. So I concentrate on the loveliness of bits and pieces of the Atlantic. My focus is narrowed, and it is intense. Is this not what I'm doing in my career and, indeed, with the cabin?

This the Last Day at the Beach: if I am Zen at all, then I must be partly Baptist as well, for I believe in the power of the water. People come down to it with all their humdrum hang-ups, foibles and anxieties.

I am a gentle spy. I discern this couple's walk, that old man's walk, I catalog their stride, the way he looks out of his face, inventing dossiers on what he does in real life. I am a camera.

They come to the water stripped down, too. And I watch the water beat their hurt and worries out of them, lull it out of them; thrash, tumble, humble, spin-dry and wring it out of them.

What man, topsy-turvied by a wave that started 2,000 miles away, can hope to retain a shred of his gray, pin-stripe, three-piece-suited vanity?

He presents himself to the healing waters, and is renewed, just as after this re/treat I am baptized with purpose, and my vision is renewed. It is time to return to the front.

Thursday, July 3, 1975 — Tonight I feel like a mud-dauber. I started chinking the spaces between the logs today, and we've come up with a strong, air-tight and insulated form of what the old-timers around here call "daubin'."

Foxfire suggests red clay, straw, water and hog's blood to make the

In Which We Chink and Daub 75

stuff bind. But I want the chinking to be gray like the logs, so I'm going with the same mortar we're using for the rockwork — the only difference being that I mix this stuff wetter than Kurt likes it for his work.

Cousin Gene Ham suggested wrapping fiberglass insulation in a chicken wire "sandwich" and stuffing this into the space between the logs first, toe-nailing this into the space, and then slapping mortar or either side.

It's a tedious process because the fiberglass has to be cut with big shears, as does the chicken wire with wire clips. It's itchy and scratchy work, but once I've nailed this insulated lathing into place, the actual chinking is fun to do.

I mix about as much mortar as I think I can do in a couple of hours, set up a log (one cut out for a door) on end for a working platform,

An up-ended log serves as a table. I've got a mess of mortar ready for chinking.

and use a small piece of plywood as my mortarboard, and with two small trowels, go at it.

The work is pleasant during this hot summer. There I am in the cool shade splatting the stuff on the lathing. I start from the bottom of the space with the mortar and work my way up. The mortar simply grabs onto the chicken wire and I finish each row at the top with a narrow rockmason's trowel. All this we invented ourselves, not many folks around who can show you how to chink a log cabin — besides, we're making this method up as we go along. Then, after one side dries in a couple of hours, do the other side and you're working against a rock-hard resistance.

Working higher up on the cabin is tedious. You have to balance the mortarboard on the ladder's top rung and then often as not you spill the whole thing down your front and onto the floor.

This chinking business takes patience and an artistic love of dabbling in gooey things. I find I seem to have a knack for this mixing and working with "mud." Maybe I'll become a potter someday.

I like to envision some future winter day when this chinking will be keeping out the cold — and we're seated around the massive rock fireplace recalling all this. We'll edge closer to the flames and chuckle about the time that Grandpa Jock almost fell off the ladder and spilled a whole mortarboard of cement down his front.

Saturday, July 5, 1975 — As the cabin goes up, it looks like I'm falling apart. I've lost so much hair now in the back that I look pretty awful; don't have any idea what it's about. I've just developed this big bald place in the back. This happened once before three years ago and it eventually all grew back.

And not only am I going bald, but I've also had to start wearing glasses again. The constant wood dust, log cabin grime, and mortar all over my hands have forced me to abandon my contact lenses. So I no longer look like the same dude who started this madness just a year ago.

Sunday, July 6, 1975 — We have just finished the biggest three-day push of the summer. Kurt returned and rocked along ever so slowly; he's now just above knee level. I mixed mortar and chinked the north wall; there are 11 spaces to chink, 22 feet long, with the cracks varying from one to four inches.

Yesterday Bill came down and helped me frame in the kitchen. We built the window frames based on the rough opening sizes of the Andersen windows I intend to buy some day.

Like a rough sketch, the framing gave me such a good idea of how the kitchen will look someday. The power of suggestion of the skeletal outline of those 2x4s is amazing to me.

Saturday, July 12, 1975 -- Things are wild at the paper: we are seriously considering buying our own press, a used Goss Community from Meterie, La. And this week we have received the first solid nibble that the *Courier-Sun* might be willing to sell out to us. Holy cow, that would be

In Which We Chink and Daub 77

Selena watches as balding dad chinks up high.

the most incredible thing that could possibly happen to a bunch of young journalistic whippersnappers. But we can't breathe a word to anyone about this deal while it's in the works.

Saturday, July 19, 1975 — Impossible. The cabin-raising was a year ago. Now the cedar shingle roof has turned a pale gray, with only the slightest hint of golden days.

The progress on the cabin is tantalizingly slow, yet it is an everlasting source of energy, peace and joy. Finishing will always be the final objective, but not the only one. I'm having a hard time explaining this to people who wonder why we've not moved in.

I've inadvertently stumbled on the kernel of my building philosophy: it's not so important how the cabin is coming... it's how *I'm* coming.

How can I explain? Just being here is a worshipful experience. I work in a state of wakeful dreaming. Like T'ai Chi, building this house is a form of moving meditation.

Over the last year the cabin has assumed an identity and a life of its own. It will not be bullied or rushed. I have had to learn to listen and pay attention to the cabin's rhythm.

One man working alone. Each new operation is its own education. I learn by degrees, new active verbs in gerund construction: joisting, flooring, rocking, chinking. One board at a time, one 4'x8' sheet of plywood at a time, one rock at a time, one trowel of mortar at a time, one day at a time.

The would-be woodbutcher learns lessons not only of patience and human endurance, but also of applied mathematics which never sank in during his formal education. If that's so, then this must be my "informal" education. I do my sums and figuring with ballpoint pen on scraps of wood.

Housebuilding 101 in the wilderness classroom, with the cabin as a hard but fair teacher, brings to mind the adage: "Tough teachers make for tough pupils."

This old barn has taught me a thing or two about problem-solving. Never before have I, one man alone in the woods, had to grapple with such an imposing task. But simply by confronting the problem, and then disassembling it, log by log, and then putting it back together piece by piece, I have gained great insight into the logical progression of problem solving.

The Zen-Quaker builds like this:

— Becalm yourself. Purge yourself of your anxiety.
— Identify the problem. Confront it and demystify it.
— Get the Big Picture in your mind's eye and assign priorities, articulating the steps.
— Then compartmentalize. Focus completely on just the one task at hand, and do that task wholeheartedly to completion.

— Move on to the next logical step. And the next.
— Completion is a function of nothing left to do.

Saturday July 26, 1975 — Salvaging is a complete religion. We've used everything on this hog but the squeal, as the old saying goes. Now we've even used the very ground the place stood on. After years of animals living therein, you can imagine the potency of that soil once the plot was bared to the sun last summer.

That's where we plopped our seeds this spring — and what a garden has sprung up. I feel like Jock and the Beanstalk, what with Swiss chard, squash, and kale practically leaping out of the ground. I have to eat spinach sandwiches just to get rid of the stuff.

Now it is late afternoon and I am sitting in the driveway up the hill above the cabin watching the sun march across the shingled roof. Behind me up the hill our resident woodthrush warbles his liquid summersong.

I've been chinking all weekend alone since Maggie and Selena have gone to Walt Disney World for the week. The chinking goes well; it is whiter than I had imagined it would be, but also everlastingly strong, which is gratifying.

Another self-discovery along the way: I may be more of a craftsman than a scholar, in spite of my college town upbringing. Chinking is an artistic endeavor, more like painting with a palette knife... and it's an archival concrete painting. And today I notice that I'm approaching the halfway mark in the cabin's chinking.

Sunday, July 27, 1975 — A threatening frogstrangler has just blown past to the north, missing Hogwild by one valley.

Ruby the Weather Dog creeps around my feet whimpering as the menacing thunder grumbles from the head of Sourwood Holler's Silver Creek. Maybe she was struck by lightning recently, a friend suggests of old Rube's new paranoia of thunder. She's eight years old and has just started to go bonkers over storms this summer. I give her the comforting she wants: we chat about the weather, I tell her there's nothing to worry about, and she apologizes with sad auburn eyes for bugging me so, but she can't help herself.

Wednesday, Aug. 27, 1975 — In today's edition we announced that we are buying the competition paper. The changes at *This Week* going on right now are mind-boggling: purchasing the *Forest City Courier-Sun*, combining the two staffs, and then installing a new press next week. This is what we had been working toward all along and I reckon "fullness of time" had reached critical mass.

Co-Editor Ron Paris said in today's paper, "The events of this week have been monumental ones for us." To put it mildly. We both know this is going to be one crazy autumn.

It is late afternoon. I am down by the creek with typewriter, trying to get some perspective on it all.

We carried straw mulch out this afternoon for the garden, and before I write I have to pick golden strands out of the Royal's keys. The sickly-sweet smell of bug repellent wards off the persistent buzzbombing of no-see-ums. An Ohio Blue-Tip explodes, and a corncob pipe responds, sending tobacco incense rising like the late afternoon thunderheads straight up into the umbrella dogwoods overhead.

The typewriter, ordinarily an officious-sounding braggart when enclosed in four walls, sounds properly muffled by choruses of crickets and cicadas, and by Silver Creek's nearby incessant "Shhhh —," the waters scolding the old black Royal on the picnic table.

Late August is the time of the insects. They wheeze, chirp, buzz and throb their last mighty tune. Butterflies, sensing the end of some golden time, grow emboldened. Fritillaries bask on hearth rocks, swallowtails show off on the iron-colored Joe Pye weed — and one friendly zephyr has decided to be my mascot and follows me around the log cabin like a spotted pup. They all know a change is coming.

My sourwoods are all individuals, some still in high summer green, others brazenly splashed with red wine plumage. A light breeze stirs the Holler, sending an early-turning flame-yellow tulip poplar leaf on a plunging sky-dive.

The first falling leaf of the year sends an oddly thrilling sensation of impending change through me; an eagerness to be away from summer's mugginess and on the with urgent business of autumn. And the business at hand is very urgent indeed. At *This Week* we are leaving an old phase behind and entering a new era of power and independence.

We worked for this moment for seven years, and now that it's within our grasp, the magnitude of its actuality is almost more than we can comprehend. This fall I imagine I'll need as never before the creative, regenerative power and sustaining energy that Hogwild gives me.

Saturday, Aug. 30, 1975 — Reporter friend Lew Powell of the Charlotte Observer interviewed Maggie and Jeanette for a piece he did on how women of the Carolinas are dealing with women's liberation and their own private lives.

The headline over Jeanette's story read: "Ski Bum Turns Earth Mother." Lew did a good job catching Jeanette, who in many was is the "earth mother of Hogwild farm." She told him that when they were living in Colorado, she told Bill "If you ever want to move back to Rutherford County, I'll leave you." But she changed, and of Hogwild she said, "This has always been Bill's dream house; he spotted it a long time ago. I begged him not to bring me out here. I didn't want to love it as much as he did.

"I'm peaceful now. Marriage had a soothing effect on me. I feel like I have a purpose in raising a family. We couldn't have it this good anywhere else. Bill's brothers all live around here, and we have a good, down-home family thing.

"Sometimes I feel confined even here. But usually I just look around and think 'all this laaaand and nobody else around!"

Jeanette, 26 now, told Lew, "I'm the same beebopper I've always been. It seems like you get to a plateau you're happy with and then you spend the rest of your life working to stay there.

"Sometimes I picture myself at 55, working in the garden, wearing a bonnet, with a bunch of children trailing along behind me."

When it came her turn, Maggie told Lew she tried at first to be a traditional stay-at-home wife but that "something else in me had to have a chance to breathe. I didn't understand it at first, why I felt so frustrated. I loved my home and my child, but I resented the he's-the-Papa, I'm-the-Mama thing."

Maggie told Lew about her career as a jazz singer, folk musician, and artist-in-residence with the schools. She said, "Jock just takes over on weekends. He's responsible for Selena and the house. He rejoices in my fulfillment and is not jealous at all."

But she concluded, "If I had to chose between my marriage or my independence, I couldn't decide. They're both important to my happiness."

8

The Cabin Grows as Friends Pitch In

*"The ornaments of a house
are the friends that frequent it."*
— Ralph Waldo Emerson

Wednesday, Sept. 24, 1975 — I feel as if this weekend was the first time I'd seen my family in 20 days.

Over the last three weeks, I have had scarcely a moment to write in this journal — let alone work on the cabin. We installed our newspaper press, bought the competition newspaper and integrated the two staffs, moved all their stuff over to our new offices, and are putting out our own papers and putting in some incredible hours.

But this weekend was beautiful out on the land, and the first time I'd had to work on the long-suffering cabin since early August. Sourwood Holler was resplendent in crimson; a battalion of Beefeaters in gray breeches and hardy burgundy coats stand watch as fall slips unobtrusively down through the foothills.

Overhead, the high cirrus clouds are sharply defined

Losing hair but gaining ground: I take over the chimney.

against the art deco blue sky. No hazy umbrella of muggy air. Instead, autumn washes the cabinman with invigorating breezes -- blowing all manner of things about — including shards of fiberglass insulation. Ouch.

The Cabin Grows as Friends Pitch In

Maggie is flanked by clogging buddies Dan Mason and Danielle Withrow

We're using the stuff for our insulated chinking, and it's nasty business. You've got to wear a Sherman tank or at least overalls, long sleeves and a bee-bonnet to work with the pink stuff. But I must be allergic to it; I break out in bumps just thinking about it.

It is slow and toilsome work. At the rate we're going, we'll be all chinked by Valentine's '76. As we get higher up the walls, the work becomes progressively tougher. Saturday, Maggie and I were assisted by friend Deborah Compton, who concentrated all day on one corner. It was her first time at chinking, but she still couldn't resist signing her name in the mortar when she was done. Ain't that just like a Gemini?

As the day wore on, and we got over our heads, so to speak, the only way to get where I really needed to work next was to stand in the chimney. Actually what I did was to stand in the firebox, but it still must have looked pretty silly. We got pretty punchy about chinking, making up bad jokes about it: I feel like I've got mortar on the brain. Or imagine a log cabin construction company with the framed company motto hanging on the wall: "Chink."

Ah, the trowels and tribulations of chinker.

I may really be out of my mind, for today I started rocking the chimney all by myself. I've about concluded that Kurt isn't coming back, and since I was working around the chimney with mortar, I wondered to myself: well, what about this rock from the pile going here on the chimney. And much to my joy, it fit. So I kept at for a couple of hours and laid several nice rocks and repaired several of Kurt's falling rocks. Hey, this is fun!

Sunday, Sept. 28, 1975 — Another great weekend at the Holler where I feel like I'm being regenerated and restored after the power drain of the 19-hour pressday.

This one man's family had a chinking party today. And what a time we had. Friends Danielle Withrow and her new husband, Dan, along with Deborah Compton, pitched in for a day of progress and silliness.

And then we settle down and have a chinking party.

I decided that the aggregation should be called "Little Maggie and the 3-Ds." While they chinked, I mixed and forked mortar — prompting Dan to suggest that I was the mortar-forker.

More bad chinking jokes continued. "I'm chinkin'!" Dan declared, striking an affected "Hee-Haw" pose. "And I'm grinnin'!" responded Danielle. Then, without another word, they both threw down their trowels and commenced to buckdance on the cabin floor. Maggie joined in with the footy tattoo, and the three clogged around the place, laughing and stomping, raising a fine mist of mortar dust that wafted out between the logs. The foot-stomping good country weekend made me all but forget about the 19-hour Tuesday behind me, and the one coming up.

So ends an incredibly busy and lovely September.

Log Cabin Gothic. After the party, the ensemble poses for a serious portrait.

Saturday, Oct. 18, 1975 — I have become a disciple of the "Sleepy Smart School of Chinking." Have no idea who Sleepy is — only that Robert Hensley down at Hills Hardware has enlightened me about Mr. Smart's method for making chinking both warm and permanent.

Here's the recipe: after mixing half a bag of Portland Cement (Type One) with 15 shovelfuls of "Columbia white" sand, and adding appropriate water 'til it's stiff, you throw in a smidgen of something called

Acrylic Concrete Binder (made by Bonsal). A smidgen in this case is about half a cup.

The result is mortar that resembles concrete "Cool Whip." It sort of plasticizes the chink, making it easier to spread to the point that we feel like we're frosting the cabin. I'm told this space-age chinking will resist cracking, expansion and shrinking — the latter two being a chronic log cabin problem with winter's temperature extremes.

Sunday, October 19, 1975 — I have to wonder about closure. When does the builder ever get to stop and say now, it's complete. Like the live thing that it is, this log cabin seems to grow more than it is "built." Maybe it will always be growing, and thus never "done."

If so, then the cabin is like child whose growth we mark on the side of the bathroom door with pencil notations of height coinciding with events: four-foot, five-inches at kindergarten graduation; the bathroom doorframe as vertical diary.

This week we mark the "door" with a fresh growth entry. My 90-year-old grandmother not only drove the 200 miles from Chapel Hill — which in itself says something about the regal woman's abilities — but she also spent Saturday with us doing what we would have been doing at the cabin anyway: chinking.

Grammy Lionne Rush (Vassar '05, DAR, Republican Women's Club, University librarian and curator of the Rare Book Room) is a doughty old queen, and it never occurred to me she might have anything in common with our woodsy venture. So we were pleasantly surprised to find out that she and Granddad Rush had had a similar experience when they were young parents back in the '20s, reconstructing a couple of log cabins within a loose community of friends in Indiana. So she knew whereof she chinked.

It was a poignant time for her. The cabin, its woodland setting, and the chinking brought back memories of those happy Indiana log cabin days, of her three girls as young children, and of her husband, Charles, taking particular pride in designing the rock chimney so that it would draw just right.

She took real joy in helping us build our own home. And it was good for us working together on such a permanent edifice. We all felt a sensation of connectedness — past to present, present to future.

I loved the picture I took of her chinking — she with trowel in hand and a happy expression. She's handling the tools with as much confidence and dainty flair as if she's serving tea to the University women's circle.

Who says you can't chink at 90? Grandmother Lionne Rush of Chapel Hill shows us how.

I'll always remember Gram chinking that corner of the kitchen. I'm reminded again how that when each person helps us on this cabin, their offering of presence, time and work is memorialized in wood, rock, and mortar.

Sunday, Nov. 2, 1975 — October relinquishes the calendrical baton to dashing November and the 1975 Relays move to the 11th runner.

It is a smooth transition. The pass goes hardly noticed. There is more stir made over the daylight saving time switch and earlier sundown than there is fanfare welcoming the month of Scorpios and Thanksgiving. Even the weather seems to have forgotten that winter is supposed to be on the way.

The log cabinman worked shirtless this weekend, rocking and chinking in the Tar Heel Indian Summer. I observe fall's welcome arrival: wasps who have taken up residence in the airy kitchen (framed in, but no windows) zoom and dive-bomb like tiny Stukas.

This autumn the leaves just crumpled and gave up the ghost without much attempt at color — save the sourwoods which went out in a blaze of auburn glory.

And now, with the leaves all down, we can look out from the cabin's perch on the knoll as if someone has dropped a green curtain on an outdoor amphitheater spread before us, revealing the creek winding its way below, the bridge leading across to the graying Soccer Field, and the bottomlands lying fallow and sere beyond the mouth of the narrow valley. For the first time since we built it last spring we can see the kid's new Playhouse from the cabin. A sense of a homestead is beginning to creep in here.

I feel like king of all I survey, and since I can abruptly witness more of my wild kingdom, my sense of proprietorship is enhanced. Sometimes I stand on the lip of the kitchen's creekside prow just gazing out and listening to the last of the wheezy crickets singing slower now, like an entomological chronometer, old November's song.

Sunday, Nov. 9, 1975 — Kurt came down today for the first time in months, and pitched in to rocking the chimney again. It's about waist-high now. Good to have him back.

Yesterday Maggie and I drove to Tryon and bought all the siding for the cabin at Ellis Kuykendall's salvage place in Tryon. The 10 and 12-inch-wide, rough-cut pine boards had grayish-blue streaks in them, because, Mr. Kuykendall allowed, they'd been cut in the dark of the moon. So we got the lot at the bargain price of $85 and hauled the load home in the back of the newspaper company van.

Sunday, Nov. 30, 1975 — While Kurt rocked along, Maggie and I had a glorious day, completely siding the kitchen with the pine boards. We worked out a system wherein Maggie measured and marked the boards while I'd saw and nail them in place.

It was so hot and sunny that I was forced to shed my shirt and work in the warming sunlight. It was a thrill to see the walls go up, enclosing the kitchen wing. Hey, this place is beginning to look like a sure-enough house!

Kurt is about one-third the way through the chimney — roughly five feet tall now — and this requires one set of scaffolding to facilitate hauling the rocks up to their resting places.

Thus another stolen day on a stray weekend turns dreams into reality. We pour ourselves into this heart pine hope chest.

I can't say exactly when it happened — when I first started thinking about this old log barn as "Home," but it has surely happened. And it's

The Cabin Grows as Friends Pitch In

not just the house itself, but where it sits, the land it occupies.

There is a spirit in the Holler that I sense no other place. The valley fairly resonates with mystery and history. The trees have souls. Selena insists that the fairies live on Slickey Mountain. And when she tells me that, I don't laugh at her. I wouldn't be the slightest bit startled if one day a graceful tulip poplar spoke to me... or if on a full moon night I caught the flitting movement of distant figures dancing in some firefly-lighted glade... or the sound of faint pipe music intermingled with Silver Creek's chorus.

Of this I'm certain: I'm a guest in the Holler's hospitable halls. Not the owner, nor the lord of the land extracting his tax and toll — but rather a privileged and devoted caretaker. The rights of this cohabitant do not exceed those of the hawks, deer, foxes, and owls who know my land more intimately than I could ever hope to.

The difference between me and the animals is that I'm aware of the land's beauty and worth, and fully cognizant of my frightening ability to destroy it. I'm the only animal in Sourwood Holler that can do that.

Someday, our log cabin home will reflect that reverence for the land. I would be most humbled and honored if someone told me it already does.

We have just sided the kitchen with salvaged pine boards bought cheap because they were cut in the wrong phase of the moon, according to the seller.

9

In Which We Move a Smokehouse and Create Another Hogwildling

> *"Luck is the residue of design."*
> — *Branch Rickey,*
> *as manager of the Brooklyn Dodgers*

> *"Chickens come home to roost"*
> — *traditional American folk saying*

Sunday, Jan. 11, 1976 — One good cabin-raising deserves another. One of our brothers of the land, Gene Ham, who participated in my cabin raising, has now joined us at Hogwild.

We have found a terrific log house for Cousin Gene Ham, Now, to move the monster...

In Which We Move a Smokehouse 91

Once we get the logs down, we pitch in to load them on the truck.

Gene, who teaches at Clinch Valley College in Wise, Va., is an old '60s classmate of Bill Byers' at Sewanee, where they discovered they were related. I started calling him "Cousin Gene" too, though he wasn't kin to me at all. But he seemed to be part of the Hogwild family from the start, even though he lived so far away, so the title fit.

After repeated visits to Hogwild, Gene was finally ensnared by its charm. We encouraged him to join us — and he couldn't resist when the Byers found a log house over in the Shiloh community about 15 miles from our land. Gene bought a couple of acres on Hogwild just above the site of my barn and the tenant house.

Down in a cove, he'd discovered an old chimney, and local lore has it that parts of my barn came from an old log cabin that once stood there — which explains why some of my beams are notched as if from a previous lifetime. "Big Daddy" in the kitchen has notches that weren't being used when I found the barn in '74, and the style of notch is dated at 1850, according to historian Eric Sloane. Also, the "Zonk" series of vertical log supports in the kitchen had the same kind of "double-seated rafter notches" which were designed for use in a horizontal position, proving they too were in a second lifetime when I found them.

Over the last four days, a slew of us have pitched in and built Gene his cabin, in much the same fashion as we did with my barn. But because of the distance that had to be covered, it was a more complex operation and it took a lot more friends.

Gene, a wiry Southern gentleman who refers to himself as "that 155-pound weakling from Tupelo," had more than his share of the action. During the take-down of the cabin, he fell off the roof and went right through the floor, a drop of at least ten feet straight down. Gene, with his characteristic dry wit, allowed that the fall "was done alone, and thus was made more bearable, since I was spared the embarrassment of it all." Then the next day he fell again — this time while he was working on top of the cabin and grabbed a rotten log for a hand-hold and the thing gave way. Gene toppled backwards onto the ground and the rotten log came with him, just missing him when he landed.

No harm done; Gene was just learning the first rule of log cabin building: keep your tetanus shot current. I still maintain you've got to hurt a little with a cabin before you can really birth it fully. And if that's so, then old Gene certainly paid the price of initiation.

At the razing site Thursday, we had assembled a crew of eight to 12 folks who toted logs and heaved beams onto the flatbed truck of our generous neighbor, Hugh Whitesides. Hugh is a great bear of a man who farms and raises pigs. Hogwild borders him on the west, and you'd think a native Rutherford Countian might look askance at all us long-hairs building cabins back in the woods — but no, Hugh has been befriended us all, and in turn, he is invited to parties, gatherings and musical shindigs.

The rest of the workers included Hogwilders Bill and Jeanette, along with brothers Stuart and Burwell; their friend Richard Davis from Forest City; Jeanette's cousin Jerry from Atlanta and girlfriend Sandy; as well as Hugh's two boys, Billy and Bobby.

Once the truck was loaded, Hugh slowly drove back to Hogwild, and then carefully edged down Gene's driveway, which, like my road two years ago, wasn't prepared for a truck of that size. So, after getting mired to the axles, the truck was abandoned and

Cousin Gene looks pleased and he ought to be.

In Which We Move a Smokehouse

Gene's place goes up quickly. By this time, we had six log barns or old cabins on Hogwild; we were vets at this business.

we carried the logs in by tractor and a trailer borrowed from the Byers' gas company.

By Sunday, a rock pier foundation had been expertly crafted with some help by Hogwild neighbor Bill McCullough, who is not only a painter but also a gifted rockmason. We were ready for another cabin-raising, and based on what we'd learned from doing my place, we were practically veterans. Of that original crew, six out of the seven were here for this one.

This was the best way I could express my thanks to Cousin Gene for his help on my place. A cabin-raising is a gift; and like the best gifts, its value and beauty lies in sharing. Something very special happens when you're 15 feet up, clinging to a log wall with one arm, muscling a mammoth log upwards with all your strength, and you're giving your gut-busting best for someone else's house...I have come to discover that the gift works both ways.

And I don't think I was imagining this: that during these cabin-raisings I sense us all getting off on how historically appropriate such a building party is — especially in this, our Bicentennial year.

The success of such an event is dependent on every man, woman, and child there (and we used all three). Every time eight people put their backs into a log, and it slid into its well-worn half dovetail notch, there would follow a communal sigh or satisfied bursts of alright's and

94 *Hogwild*

yeah!'s and now, what's next?

As Sunday wore on, and the log structure rose higher, Sweatmore Construction Company had added friends Jerry Green, Erwin Johnston, Toni Shell and Peggy Shearon. We needed every ounce of muscle we could muster.

We have learned not to underestimate these 500-pound logs. At any given time, a slipped foot, a missed grasp on a rope or log could result in disaster.

Perhaps it is this added danger factor that makes a cabin-raising race for its climactic ending. The higher we go, and the tougher each log becomes, the more the excitement builds.

In Sunday's faint drizzle, we were treated to a special country event

Three-quarters up, we pause for a Sunday feast.

Oh no, not another log barn. No problem, we say. We can handle it.

Wildman celebrates roof coming off the Pea Ridge Smokehouse.

when Jeanette and Hugh Whitesides' wife, Bernice, whipped up a picnic for the cabin-raisers. We had about three more courses of logs to go, and those logs which were lying about on the ground now became impromptu tables and benches for an old-timey Sunday feast. A bushed crowd of workers flopped around on the forest floor to chow down.

"Don't you know," exulted Jeanette, "this is the real thing."

I could see in Gene Ham's face the same sort of amazement I felt at my cabin-raising. "Yesterday I never would have believed it," he mused. "And today, it looks like it just grew here hundreds of years ago."

Friday, Jan. 16, 1976 — I have a bad case of cabin fever... and a couple of cracked ribs.

Here's the play-by-play: three years ago, while still living in the big house on Pea Ridge, Maggie and I had found a small log barn way back in the woods. For $125 we decided to buy what turned out to be an old smokehouse. The remains of a fairly decent road still led to the old homesite. So yesterday Maggie and I rented a small U-Haul truck and eased our way into the site, figuring that we could take down that 12'x15' barn in an afternoon without breaking a sweat.

When will I ever learn?

Even though we had labeled it and knocked off the rotting roof

96 Hogwild

*But alas, this barn is bigger and heavier than we reckon for.
(Photo by Selena Lauterer)*

And it's all that Maggie and I can do just to get it down and onto the truck.

earlier, what appeared to be a "cute" little barn, turned out to be a bear.

Maggie and I strained and gutted it out, first taking down the barn and then hefting the big logs into the truck. The logs were simply heavier than I had estimated. Is there a Murphy's Law of Building Log Cabin? If so, it must go something like this: "The given weight of a any log increases incrementally and proportionately by the square root of how many people are there to lift it."

In other words: you never have too much muscle. Well, we just flat busted a gut — that's all there is to it. The pictures that Selena took of Maggie working as my helpmate are infinitely revealing; she doesn't even look like herself. And during the work she broke down and cried at one point, "I'm just not strong enough."

So that made me strain to Herculean limits — and when we found that the bottom four logs were chestnut and heavier than all the rest of the pine logs, it took even more muscle than before.

But in the end, we did get them loaded on the truck, and we lumbered back across Rutherford County to Hogwild, all nasty and grimy, where at Jeanette and Bills' we were fed and wined while we regaled them with stories of our brute strength. We were sitting there eating and drinking and trying to get over it when one of the Byers brothers (I don't remember who it was) chortled, "Hey! we've got all these people, why not put up the barn right now!"

It was one of those weird nights — warm for January, a full moon, plenty of spirits. So why not put the barn up tonight?

There was a lovely full moon lighting the valley like a huge bluish spotlight, and the mild January temperature was in our favor, and since we were high as kites from all manner of things including the day's work, I figured — why not?

So the five of us jumped up from the table and trooped down to the Holler. The crew included the three Byers brothers plus Keith "Too-Tall" Harrill, who at 6 feet, 10 inches, came in mighty handy.

Earlier I had built a minimalist block corner foundation. And onto this we started laying the logs. Nearby a fire was started and someone had some Boone's Farm and I guess we were all pretty zapped — for on the fourth course around, we inadvertently toppled over one of the foundation piers and the whole thing fell with a crunch.

Not to be deterred, I piled the blocks back and we went at it again, this time with five hefty dudes making the work go fast. It was such craziness, putting up the smokehouse by moonlight while we were so buzzed.

I was inside of the frame at the very apex of each corner, and then guys were handing me up log-ends to lift and place into the grooves — when we got up to the top and the only person who could heave the log up was "Too Tall" Harrill...

...And the only way I could reach the log he was handing me, was to lean over the top logs and put all my weight on the edge of my rib cage...

I was hanging off the steepest corner, which must have been 10 feet off the ground, with big Keith stretching all of his immense frame while he handed a big 15-foot-long log up to me. When he did this, and I grabbed the log, I remember feeling something give inside me, but I was having too much fun at the time to wonder what it might be. Using my rib as a fulcrum, I gained another log barn and lost a rib.

That's what the doc says. So I've got to wear a brace for a couple of weeks, and it really hurts when I laugh, and that's pretty silly because the brace is really like a girdle and I've about heard my fill of "my girdle is killing me" jokes.

Wednesday, Feb. 11, 1976 — The other day I hosted a class of first graders out here at the cabin. It was a part of their study of early American culture in connection with the Bicentennial celebration.

I was trying to impress the 6-year-olds with the rough-hewn quality of pioneer life as they toured the place, showing them the tools of the American backwoodsman: the froe, the maul, the adze, the auger and the broadax... until we arrived at the kitchen where I was giving a lecture on how the old builders pegged their log homes together. I took pains to stress how everything I was using at my reconstruction was likewise imbued with age and antiquity — until one little boy interrupted loudly, "There's something that's not old!" pointing at an empty

The tools of log cabin building: clockwise from bottom, froe, auger, pegs, adze, maul, ax, broad-ax and framing hatchet.

Coke bottle. So much for faithful reconstructing of the pioneer life.

Saturday, Feb. 21, 1976 — A light rain falls. I am seated in the kitchen (framed, sided, but still windowless). After a sip of water from the creek, I observe the day's work — roofing the new log smokehouse.

Indispensable handyman Joe Ryan came by playing his penny whistle, and ended up helping.

The cracked rib has healed slowly so I've worked at an easy pace this month. I constructed the roof by first using two 18-foot-long pine poles left over from my original barn for the top sills. Then I tried my hand at notching and pegging for the first time. Pegged the new sill into the old top log, and notched a place to seat the rafters.

Last week, two zanies, Andy Morgan and Sam McMillan (both public school artists-in-residence) came out and helped hoist the rafters I'd built out of new lumber. Today I planked the roof with the old rough-cut pine boards salvaged from the original log barn almost two summers ago.

This afternoon I thought I was hearing things: I could swear it was a penny whistle playing a Scottish hornpipe or an Irish jig. But because the spring peepers have started their soprano spring song today down in the marsh, I couldn't be sure at first.

Stopping my work, I cocked my head towards the sound. It was a penny whistle all right. And as the music grew louder, I realized the music maker was coming right down into the cove.

Then out of the trees steps Joe Ryan, a neighborhood handyman, old-time music maker and transplanted New Englander, piping gayly on a little silver tin whistle. It was a positively haunting sound in the Holler. Joe climbed up on the rafters and played me a tune as I resumed my work.

Joe, who takes Saturdays off because he's a Seventh Day Adventist,

The Byers get together and make a purchase so they can do some serious farming. Jeanette waves as Burwell drives and Bruce rides.

Neighbor friends Max and Michelle are married, and we give them a party at the Big House, beginning with Joe Ryan on the bagpipes...

...with dancing to follow.

elected to help me anyway, and together we finished the uphill side of the roof.

As to what this building will be, it's anybody's guess; a guest cottage, Maggie's studio? It really doesn't matter. I find it satisfying to have another log building on the land, so that while I'm working on one, I can gaze at another... sort of a mirror effect. For now, we just call it the Smokehouse in honor of its first lifetime.

Sunday, March 1, 1976 — We are blessed. Another child will bear our name, our life, our style, our heritage as Americans and Hogwildians.

This was not an accident,

but neither was it planned for just now. Rather, we think it was just supposed to happen. But we couldn't make up our minds when it would be convenient to have a second child. We believe the Smokehouse made that decision for us.

If memory serves, we were being natural this winter and using "rhythm..." Well, as best as we can figure it, all the heavy lifting and heave-ho'ing that poor Maggie did back there helping on Gene's cabin one weekend and razing the Smokehouse the next simply threw her cycle into a revolt. And the following week we went to the annual North Carolina Press Association convention and I reckon while "celebrating" any number of things including seven more press awards, I sired Liza/ Jon. Whoopee I say.

She/he is scheduled to be born in early October, from all reliable guesses.

Sunday, March 8, 1976 — We durn near burned up Hogwild today.

We were burning brush near Selena's Playhouse, and the fire got away from us. Fed by gusting March winds, what had started out as a tame little blaze exploded as it hit dry pine, old grass, and dead vines. It was as if someone had thrown gas on the fire.

Maggie beat on the flames with her hoe, and I hollered, "I'll get buckets!" and ran scrambling licketysplit across the creek up to the cabin to get all the buckets I could gather — grabbed up four big heavy-duty buckets I've been using to haul junk rock and mortar, and dashed back down the hill to the creek, filling the buckets and running at top speed the 25 yards between the creek and the fire.

By this time it was 20 feet wide and leaping 10 feet into the trees at the base of Slickey Mountain. I threw in the first two buckets of water and it didn't seem to do a damn bit of good. I could just see us burning up the whole 300 acres by our carelessness.

Maggie, in spite of her pregnancy, had just toted two buckets from the creek and while she threw that on the fire, I ran back for another load — and when I did, in my haste and panic, knocked down poor Selena. The little girl was standing in the middle of the narrow path, paralyzed by terrible spectacle she was witnessing. "Get outa' th' way!" I bellowed — but she didn't move, and I couldn't stop my forward momentum. One of the buckets bowled her over backwards.

I could see she wasn't seriously hurt. I felt awful — but I'd have to take care of her later; right now the emergency at hand had to be dealt with — and for a minute there I thought we were going to lose it. Two more racing trips back and forth from creek to fire with the big and impossibly heavy buckets sloshing with water. It occurred to me in one of those silly mental parenthetical asides that I, and especially Maggie, would never be able to tote so much weight were it not for all the adrenaline pumping.

The brain does funny things in crisis. My mind's eye snapped something of a photographic image of Maggie in her big baggy overalls and brogans, her long black hair tied back in a red bandanna. She's got a bucket in either hand, she's biting her lower lip in concentration, and she's taking these big loping strides, walk-running sort of like a Chinese coolie with all that weight. She was flat-out busting it.

Meanwhile I was trying to figure out how to get word to the Green Hill Volunteer Fire Department, and how on earth they were going to get a fire truck down in the Holler, much less across the creek — and how there was nothing between here and Clark Road but 300 acres of

On the deckside, Maggie chinks...

woods... all the while, Selena is sitting on the ground bawling her eyes out — and Maggie and I are yelling instructions and encouragement to each other over the angry roar and frightening crackle of the fire.

Rather than fear, I was awash in the strangest emotions, guilt over knocking the kid down, guilt in advance for burning up Hogwild, anger at myself for being so stupid as to let a fire get away from us, angry for Selena's being in the way, angry at myself for colliding with her, angry at her for being no help, angry because we were fighting for our land. I wouldn't let this thing happen to us. All that guilt, anger and sense of responsibility translated into superhuman energy. And I reckoned Maggie was feeling some of those same emotions.

After the fourth trip to the creek and back, we saw that the battle had turned in our favor — and this only redoubled our efforts. Another round with the buckets and we were definitely winning. We cornered the blaze as if it was a vicious wounded beast and beat it into submission with Silver Creek's saving grace.

And then, suddenly, it seemed to be over. We had won. While the coals still required a lot more dousing, we knew we had triumphed. And the first thing we did was to get all in a family group hug with Selena in the middle of the two spent firefighters. I told I was sorry over and over. I'm sure she didn't understand. I think we were pretty lucky she wasn't hurt.

As for the land, it was left with an ugly black scar and three pine trees blackened 20 feet up their trunks. It was so close; it could have gone either way, and would certainly been a major disaster had Silver Creek not been so close by.

And a strange thing, until now as I write this, it never occurred to me that the cabin was in danger, that the fire might leap the creek. But indeed it might have.

Sunday, April 11, 1976 — Selena turned seven today, and we staged a gala event at Hogwild. I bet we were the first drama ensemble to perform an improvised version of the hit musical *Oliver* in a log cabin.

The event was totally Selena's idea, one she has been cooking on for some time. Of course, as producer and birthday girl, she cast herself in the starring role. Selena's friends were Fagin's "boys."

Artist-in-Residence Andy Morgan played mean old Mr. Bumble, Maggie was Mrs. Bumble, Artist-in-Residence Deborah Compton played Nancy, Selena's little red-headed friend, Alice Roberts, was the Artful Dodger, and yours truly took pictures and doubled as Fagin.

We brought a kiddie record player out to the cabin, and ran an extension cord to the temporary power box mounted on the big tulip poplar outside the front door, and stuck the *Oliver* soundtrack on the turntable and did the show in mime.

Cleared the construction debris from the floor of the living room, we had ourselves a rollicking good play, with children romping around the old cabin, or being chased by Mr. Bumble (whose improvised costume consisted of an Army greatcoat, a blanket for padding, a broom from a staff, and a yellow plastic bucket for his helmet).

Can't imagine what we'll do next year to top *Oliver* for theatrics... just hope Selena doesn't decide she has to be Charlton Heston and have *The Ten Commandments* down by the Rocky Broad River.

Another birthday party extraordinaire: Selena plays Oliver Twist and asks "Please sir, I'd like some more." That's Andy Morgan as Mr. Bumble reacting: "WHOT?!"

In Which We Move a Smokehouse 107

Selena skips rope down Hogwild Blvd. to Bill and Jeanette's horse barn.

10

Of a Giving House and Lofty Ambitions

> *"Though this be madness, there is method in 't."*
> — Shakespeare

> *"What next, Spiderman?"*
> — Marvel Comic Books

Wednesday, May 12, 1976 — With a pen I have inadvertently written on a scrap piece of 2x4 what I think is the theme for this phase: "Ten Things At Once."

Framing. Insulation. Salvaging. Put up more board and batten. Get batten. Staining. Get loft beams sawed. Put them up. Will need help for that. Wiring? What about getting windows? Build a front porch; design same. I need eight hands like one of those Hindu gods.

Because the more involved with building I get, the more diverse it all becomes, each project is its own world apart, and yet it is all somehow intertwined under the same roof.

So instead I take time out to go sit by the creek this gorgeous spring day and reflect on it all. With apologies to Annie Dillard, I am a Pilgrim at Silver Creek:

I say pilgrim because I come down here like worshiper to a natural shrine. It is just below the cabin and so close — and yet a different world; a green chapel, the rock fireplace its altar.

I can come here and feel a certain relief from the taskness that burdens me just up the hill, confronting me, confounding me, and at times overwhelming me; a gigantic challenge of my own making.

I take stock.

Maggie is huge with LizaJon, as we have come to call the growing contraction within who'll soon be with/out. Eliza the girl, Jonathan the boy. Either or both are welcome.

A wise old turtle surveys his creek world.

Other things a'bornin': a year after we started chinking, we have finally finished.

I bought rough-cut 1x4s and I hired Joe Ryan to rip them into 1x2 battens to cover the exterior siding. I've nailed these up and we've stained the siding with bleaching stain to protect and speed the "graying" of the board and batten to match the old logs' ancient natural gray.

Bill Byers helped me frame Selena's room and the bathroom on the north wing, and in Selena's room I'm using a large plate glass window salvaged for free from the newspaper office.

Enough, I cry to myself.

The leaves are out, shielding me from the reminder of all I have yet left to do. I take off my shirt, kick off my shoes and step into the creek. Icy fingers slip fudge-mushy goo between my bare toes. All about me Silver Creek's hushaby song fills the sun-dappled afternoon with its singular symphony. I stand breathlessly still, a rock in the creek's rush; only my eyes move, argus-like; they see:

Stiff-stalked cattails dancing wildly in a sunny reflection. A solitary deer track signing the sand bank. Tiny shelled periwinkles bulldoze miniature roadmaps in the silt of the creek bottom. A clutch of fragile blue "forget-me-nots" — each blossom no bigger than an aspirin, dot the mossy bank. A wise old box turtle with rings and wrinkles to testify to his age, glares at the creekworld like an amphibious Ahab.

Life is all about me. A nondescript smallish snake swims away with energetic S-curves to the creekbank's dark haven. A spotted leopard frog jettisons himself from his launching pad. A brown salamander, slippery and slick as the neighboring creek pebbles he looks like, wriggles beneath the surface.

Woodsflowers vie for my attention. I imagine blooming clumps of mountain laurel to be pale pink strawberry ice cream cones. I catalog the modest violet, the daisy flea bane, false and true Solomon's seal, the Clinton's lily, pinxter, flashy splashes of flame azalea, purple wild iris, growths of voluptuous pink lady slipper, and my favorite — the sweet bush, which fills the coves and hollers with a pungent spring perfume.

As if stalking the very creek itself, I inch my way upstream. I am her greatest admirer. Even during a recent dry spell, the creek's spring-fed level remained reassuringly constant. This wellspring gives me cause enough for reverence.

The first thing up in the spring, a snow-white bloodroot.

I take care not to step on the thousands of fresh-water snails that line the rock-lipped edge. Seen from above, the periwinkles look like so many weekend sun-worshipers at some summer-clogged beach.

The newest green leaves of the broad basswood and downy-soft beech ovals stretch out over the water. Lowly liverwort clings to wet cave rocks. Butterflies — zephyrs, tiger and black swallowtails — flit among the shadows with a thousand other skimmers and wheezers. I stop to examine the story left by a raccoon come to the creek's sandy bank to clean his supper.

A single teacup boat — a solitary mountain laurel flower, is flung

Of a Giving House 111

into the torrent and dodges in and among the current, tumbling downstream past succulent mossy banks, fat-faced communities of clematis, shiny-leafed galax in bloom and over granite bedrocks flecked with fool's gold.

But no matter how lovely or spectacular the animal or plant life — it is always the water itself that draws my attention. Each waterfall deserves its own celebration: there is the constant contrasting play of turbulence against solidarity, of impermanence against permanence. The roiling liquid is ever changing, ever alike, constantly replacing itself, rushing on in tiny parts — yet the form appears to stay the same. The water asks nothing but gives all. So that is what the Tao means: "The highest good is like water...."

As I walk up the creek I hear her music; each falls has its own note... here a reedy tenor, there a throaty baritone, or here a pure soprano.

Virtuoso solos, duets and quartets can be enjoyed just by changing positions — as if an audience member has the liberty of strolling about on stage amidst a great chorus in performance.

I return downstream to the picnic table by the little pool where the creek flattens out and is less vocal. Here she whispers to me her secrets. Sweet silver secrets.

Fern fronds unfurl.

I step out of the creek, baptized. I leave the chapel-roofed dogwoods and stride up the hill to pick up my hammer. Am I not renewed?

Thursday, June 10, 1976 — I see we're coming up on the second anniversary of this a/mazing project. And what a growing two years it's been. "The house that Jock built" was supposed to have been done in three months.... Now we've hunkered down and put our shoulders to the task. Knowing as much as I do about mountain climbing and house building, I have come to appreciate the psychological danger of anticipating the

112 *Hogwild*

Framing and siding the north wing, the bathroom side.

An Organic Gardening Club meeting down at the Big House.

false summit. So I try not to look for the horizon. We comfort ourselves with small triumphs along the way.

This week we finished putting up the exterior siding on the north wing, which took some real doing since the boards had to work their way around the beautiful half dovetail notches which I wanted to display.

As we were putting up the last boards this afternoon, a summer thunderhead rumbled threateningly from the east over Slickey Mountain and quail called to each other, "bob-WHITE!" from their shady hideouts on Laurel Mountain.

I have also dug out the shallow side of the bathroom and built a trap door so we can finally subfloor that room, the last room in the house to get floored. The crawl space under the basement was so shallow that for the longest time I couldn't figure out how to handle the problem.

I'm hoping we're on the brink of another phase with the milling of big beams for the loft. I have contracted with sawyer Ed Rhodes of nearby Cove Road to mill me eight 3x6-inch, 18-foot-long pine beams that will stretch across the width of the living room to support an upstairs bedroom for me and Maggie. It will have closets, a balcony overlooking the living room below, and large windows in either gable.

The beams were to be ready today, but so far, haven't been delivered. Andy Morgan, my red-headed clown-in-residence buddy, has agreed to help when it comes time.

Another surprise: when we opted for a factory-made circular stairway of oak, the price came back at almost $1,200 -- well beyond our financial reach.

So we said, "trust and obey," and sure enough, once we relaxed, things worked out.

During the Organic Garden Club picnic at Hogwild last week, one of Joe Ryan's friends from his church, Carlton Mason, heard me moaning about the stairs, and he responded, "Heck, I can build you that -- how's walnut?" And he wants to do it for just the price of the materials and welcomes my help. How about that for being neighborly?

Friday, June 11, 1976 — It is indeed a special day. Happy second anniversary. Two years since we embarked on this singularly sane journey in wood.

Tonight I return alone to Sourwood Holler's loveliness, after finding that the sawmill up in Shingle Holler has busted so the beams can't be sawed.

But no matter. The beams will be ready in the Fullness of Time. And I've plenty to do....

Tonight as I sit in the gathering dusk of this log home, I'm awestruck by the simple beauty. Imagine this: a drive-in movie picture screen, and seen in wide angle are a pair of feet stretched away from the camera's view... and in the foreground are white athletic socks, red mud-stained "garden boots" and baggy size 44 overalls rolled up to the knees to catch the evening breeze.

I sit on the picnic table bench in the kitchen, feet out of the big window with its low sill, watching the fairy lights of fireflies dance in the gloom. Away from me, the trees of the Soccer Field march silently like giants advancing, their silhouettes indistinct and mysteriously fuzzy.

A pipe. Lazy smoke eddies over my head. From an old snuff glass I got at a flea market I have a shot of quieting pink Chablis. Andy's birthday wine. Thanks, kid.

Ruby the Weather Dog comes in and checks to see if I'm still here and haven't gone and left her. She thrusts her noble head on top of my thigh and loves me with eyes that speak of deep, doggy devotion. "Roooby," I croon.

Silver Creek rushes constantly with its singsong voice. A whippoor-will breaks the June evening with a persistent plaint.

In the east, where Slickey Mountain comes down to nuzzle at the nape of Laurel Mountain, a lighter glow appears, promising a full moon later on.

I reflect on the day's progress, Maggie and I working all day here — she staining the exterior of the kitchen while I framed the last section of the kitchen facing the deck. I built a small window space and hung an old door salvaged from the tenant house up the hill. I believe I will do some major salvage work up there this month. The tenant house, which was part of the original deal, is mine for the taking -- or as much as I can use. Other parts of it will go to Gene Ham and Bill's brothers -- so in the end, every bit of the old house will be recycled in a new house somewhere on the land.

Getting too dark to write now.

Sunday, June 13, 1976 — D-Day, or maybe "B" (for bald) Day would be better. Sitting there in church today, I just decided, I'd had it with five years of falling hair and looking splotchy.

I marched home from church, and after announcing my intentions, shut myself in the bathroom and had at it — first with scissors, and then with razor. I was going to shave my head or be durned. About one-third of the way through, I almost lost my gumption, but by then I figured it was too late — so I went on and finished the messy job.

Well, it looks terrible, but it feels right. Without any sun on my scalp I look like a new convert to a Hari Krishna sect, or worse.

Poor Selena. When the seven-year-old saw me, she ran out of the house shrieking, "I'm never coming home again!"

Why is this man smiling.

I can't say as I really blame her. I do look worse than a raw GI shorn of his locks at basic training. And it's already tiresome being called "Kojak" by the local yokels.

But if the shoe fits....

Maggie is amazing. She loved me with my hair falling out, and now she loves me looking like Yul Brynner. I don't think it much matters to her what I look like. But for me, I can see right now this is going to take some getting used to. The following exchange is a good example in Humility Training 101.

Bill and Jeanette's 5-year-old, Hannah, ogling the shaven sphere: "Boy, do you look crazy!"

Curly: (going along, good-humoredly) "Oh yeah?"

Hannah: "Yeah! You oughta' go home and look in a MIR-row!"

Saturday, June 26, 1976 — Haying must be the hottest thing under the sun. When we all heard that farmer Hugh Whitesides was needing to put up his hay, Hogwildlings mobilized and turned out in force. Nothing heroic, just neighborly — especially in light of Hugh's help last winter on Cousin Gene's cabin-raising.

Burwell, Stuart and Bill Byers, Cousin Gene and I pitched in, and through the sun-broiled afternoon we toiled over the heavy bales, loading, unloading, while Hugh's machinery chomped the flaxen-haired field into still more bundles. Chaff and grain got in hair, combined with sweat to nettle the workers as we combed the fertile Broad River bottoms.

Hogwild alert: we go help Hugh Whitesides gather his hay.

Bill loads a bale. Gene's right behind him, and Stuart awaits the truck.

And when afternoon lengthened into evening and the job was finally done, the hay all safely stowed away in an old house, then the cool river's swimming hole offered us its ultimate reward.

Sunday, June 27, 1976 — For the last two weeks, Gene and I have been working hard at salvaging the old tenant house, which has yielded up such a rich bounty that Maggie and I've taken to calling it "the Giving House," after Shel Silverstein's classic and wise cartoon book, *The Giving Tree*, all about selfless giving and consistency.

The place is chocked full of clear, knotless heart pine — called "forest pine" by the old-timers around here. I think what they mean is the loblolly pine, but it is the virgin or near virgin growth in the forests that were allowed to grow to such a height and maturity that the wood grew straight-grained and hard in a way loblollies don't (or aren't allowed to) grow in these days of forest harvesting.

This is the old-style, post and tenon construction.

The Giving House must have been a real beauty in its day — I try imagining 75 years ago when the virgin pine had been cut from those knotless giants of the Broad River basin. It seems almost a shame to dismantle the old house — for the tumbledown place hadn't shown its true identity until I (covered with bat and owl guano) discovered its classic framing.

It wasn't just the irreplaceable heart pine....("You can't buy that at the lumberyard anymore," the old-timers tell me). It wasn't the first part of the house I salvaged two years ago and found little to remark at. It was only when I got to the second part of the house that I was awestruck by what I had on my hands.

Taking down the upper rafters I came to the massive corners of the house where I found hand-hewn beams pegged together. And the massive "H" beams framing the mud-daubed chimney had been built by post and tenon method — the style of precision jointing and notching that predates and precludes the use of nails. In short, the Giving House is a masterpiece of pioneer craftsmanship.

Gene has already salvaged the roof, so I must work quickly now to get the wood in out of the weather. I go to the door and watch the rain approaching: a gray line of diaphanous gauze dropped like a scrim down from the gray cotton candied masses above.

I take off my broad-brimmed felt hat and lay down my hammer. Off with each grimy glove, then the soiled towel to wipe away rivulets of sweat and attic grime. The break gives me a chance to reflect on the old house's wonder and to watch the storm's advance.

The trees on the far side of the June-green valley quiver and then chuckle one to another as if the wind had passed there telling a good joke that makes them now bend double with laughter. An indigo bunting flits to a nearby ruined pear tree, the bird's brilliant blue plumage a gaudy splash against the chartreuse leaves and Confederate gray rain clouds.

The summer heat breaks; the storm brings a reprieve as refreshing ozone-rich cool air sweeps up the valley.

August 8 - 14, 1976 — I can't believe this: it's Tuesday and I'm not putting in 18 hours in the back shop of the paper office. Instead, the Lauterers have taken a whole week away from housebuilding and newspapering. Today is pressday back in Forest City, and the co-editor is sitting beachside with his gangly feet dangling in the surf.

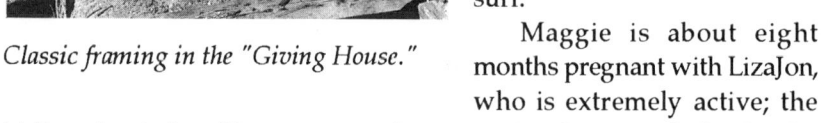

Classic framing in the "Giving House."

Maggie is about eight months pregnant with LizaJon, who is extremely active; the kid's going to be either a soccer player or a tap dancer to judge by the way she/he's kicking and hoofing around in there.

As for me, I've gotten totally used to the Yul Brynner look, and have found that I don't burn like I thought I would, and that swimming "in the bald" is an amazing, sensuous experience. I can feel more because I have more skin.

At the cabin, things have been blasting along. Two weeks ago Ed Rhodes delivered the loft beams — eight 3x6s for $132. I thought that was way too much, but when he threw in enough 1x8s left over to completely floor the loft, it seemed like a pretty good deal after all — for the beams are beautiful, green sawed loblolly pine by the looks of the logs I saw on his truck back in June.

The beams lay around until I could get the necessary help. But first Maggie and I had to decide just where we wanted the loft to go, how far out to extend it over the living room, and how close to put the beams. We finally settled on using seven beams, 19 inches apart, with a 12-foot-width extending across the 18-foot-wide room. So the loft will take up just over half of the room's ceiling space and leave a wonderful airy, open high ceiling.

I notched the supporting wall logs with chainsaw and chisel, dragged the first two beams into the cabin and manhandled them into place by myself on two separate days. The next Saturday, Andy Morgan and David Kirby (an artist who makes lovely silver jewelry) came down and donated the entire day to the lofting.

We had this marvelous rhythmic system going: I'd cut the notch while David chiseled out the block and then all three would haul the beam into place. It went quite fast, and after Maggie served a picnic lunch to the good friends, we charged on into mid-afternoon, getting the last beam up. We couldn't stop then, so we threw the flooring up there — and presto, we had a loft!

In the remaining weeks of July I spent a lot of time moving much salvaged lumber from the Giving House down to the Holler. On the last load, with the "Obie," a beat-up 1957 VW bus (named for Objet d'Art), crammed to the gills with wood, the brakes gave out coming down the steep incline to the cabin... I just barely managed to stop using gears and handbrake. Whew.

I piled the flooring from the Giving House in the living room along with the 4x5 beams, put much of the paneling in Selena's room and the kitchen, and then loaded the Smokehouse chocked full of wall and ceiling boards from the Giving House.

But I guess the weight was too much for the little woebegone log barn. This is the barn put up in a high January night when we knocked the foundation over in our excitement, resulting in the bottom side of the smokehouse being six to eight inches lower. When I added all that new weight this week it must have put an additional strain on the place. Some time when we weren't there, the Smokehouse shifted downhill off its upper foundation.

Luckily, the entire weight of the barn was caught by a log "flying buttress" we had fortuitously placed on the lower side to brace the building. (Maggie's idea, I believe.) How it remains standing is still something of a miracle to me. The tallest foundation pier is tilting

The new front porch and Maggie, eight months pregnant.

precariously — and I reckon the solution is to jack up the whole shebang and re-stack the foundation.

The last week in July we built the front porch — which was a joint project perfectly suited for the would-be carpenter and his very pregnant assistant.

Of a Giving House 121

After diddling around with various elaborate roof designs, we took Gene Ham's suggestion of a shed roof with a simple angle.

For the basic 5'x8' floor frame we used the "gangplank," a pair of 2x8s long used as the walkway up to the kitchen. I leveled the whole frame on blocks flush against the cabin, and used flat rocks found up on the main road for the front two piers. The supporting posts are stout old 4x4s from the Giving House.

The rails and banisters I made from the marvelous old grayed chestnut given to us by Dr. Hyde from his Jonas Ridge place. I've been saving them for two years now. I roofed it with poplar planks left over from last year's salvaging expedition to Kuykendall's in Tryon; he'd thrown in that poplar to sweeten the deal and I couldn't figure out what to do with the stuff.

I covered the roof with 90-pound asphalt paper left over from the main roofing job two years ago. And for shingles we elected to go first-class and use the thick premium-grade hand-split cedar shingles, costing us $18 a bundle, as opposed to $6 for the lesser grade I used for the main roof.

The shingles alone cost almost forty bucks, which seems pretty steep, but it is worth it — for the porch seems to connect the cabin with the earth and makes the barn/cabin begin to look and feel like a home and a house as never before.

With pregnant Maggie (in spacious overalls) supervising and hauling the shingles up the ladder to me, I laid them on the porch roof, and later I built some crude steps to do for a while.

After talking about it for nearly a year, we have finally gone ahead and taken another plunge, ordering the first of the expensive thermopane windows: one for Selena's room, the medium sized casement for the kitchen's east side, and the dinky casement for the kitchen's west wall facing up the hill... haven't the foggiest idea how we'll pay for them.

Selena, a saucy seven now, is forever slaying me with her commentary on life. Coming back from the beach, we took time out for a tour of the USS North Carolina in Wilmington, and as I was explaining that the battleship had seen action in World War II, Selena wondered aloud, "Wow, who won that war?"

"We did. The Good Guys." I simplify history whenever possible.

"Oh," she replied thoughtfully, scanning the ship's length. "That's why it's not all banged up."

While I was working on the cabin recently, Selena and Hannah were seating in the sandpile just below the kitchen. They were playing a game they call "Teenager." This consists of taking your toy car and zooming around the sandpile roads while dramatizing the dating game. They had planned to pick up their "girls" and drive them to "the end of the world" (as they called the edge of the sandbox).

I couldn't help overhear the exchange between Selena and her buddy, five-year-old Hannah.

Selena: "C'mon, I'm gonna go pick up my girl Julie."

Hannah: "OK, I'm gonna go pick up my girl Jello."

Sticking my head out the window, I had to laugh, "My girl Jello?" But Hannah could only smile back with her June-apple cheeks and give me a happy shrug, as if to say what's wrong with a girl named Jello?

Creek sprites Selena and Hannah up Silver Creek looking for "fairies."

Selena arrived at a memorable rite of passage this week called Learning To Ride a Bike. She seemed ready for the experience this summer and I was convinced I could teach her. So the other day back at the Blue Shed in Rutherfordton we did it. I chose to teach her in town so she could learn on pavement.

In my favorite baggy size 44 faded overalls, I launched her out on the level stretch and padded barefoot along beside her, gripping the rear of the bike's seat, chattering like a rowing coach: "Peddle, peddle, peddle... think middle, sit in the middle of the bike, turn your handlebars the way you feel you're falling...peddle-peddle... c'mon, you're not peddling!"

We must have been a sight: a bald dude in clown-sized overalls barefootin' it down the road with Little Bright-Eyes teetering like a wind-up doll in the stirrups of the blue Huffy.

And on that day something magical happened — you know that feeling — when you discover the mystical missing key to a solution. All

Cousin Gene's fanciful invitation for Hannah's gala birthday party.

at once, Selena was doing it — with me running along with a hobbling gait, letting go. Really letting go. It was like holding a small bird in my hands and then having it burst into flight.

She chortled with unrestrained joy at the sensation of floating. She was on her own. She had her Wheels now, and her life was suddenly transformed from a pedestrian crawl to Transports of Joy.

She spent the next remaining hours perfecting her new-found skill. Coming in that evening for dinner, she remarked in wonder, "My room looks different... Everything looks different since I've learned to ride my bike."

And that night, after the bedtime story, Sleepyhead told us wisely, "Riding a bike is like life. You've just got to go on peddling — no matter what."

Wednesday, Aug. 25, 1976 — This past weekend I spent salvaging more good stuff out of the Giving House, the wonderful old tenant house that has been so generous to us all.

I have come to believe sincerely that buildings have souls, and emotions like restraint, hostility, coldness, or on the other hand, graciousness and benevolence. This Giving House has given of itself so well, in spite of being dilapidated and weathered. A magnanimous spirit dwells herein. And maybe this is because she knows all her vital organs were being recycled into new houses: Cousin Gene got the roof, Burwell and Stuart have felt her generosity too, but I'm getting the lion's share of flooring, wall boards, floor joints, hewn sills, at least five doors and frames, and beautifully weathered gray exterior siding.

Two years is a long time to be undressing a house — a long time to be building another — it forces the salvager-builder into an intimate relationship with his buildings; one coming down as another goes up. Board by board I dismantle the Giving House and mantle the cabin. Each board must be carefully pried loose from its setting. If it's tongue and groove, which most if this is, then this operation takes further care that I don't split out the tongue or groove which I'll need to replace. Then each board must be de-nailed, steel-brushed, and toted either by hand or by Obie the VW workbus down the trail to the cabin for dry storage.

This weekend I came to the point where I realized the next logical step was to salvage the stairway. As usual, I was working alone, so I tackled the problem head-on, starting at the top by taking off the stair treads. Then, about three treads down, I wondered aloud — why can't I just knock the stairs down intact and save taking the thing apart in fifty jillion pieces?

So I whacked the supporting studs out of the way and got underneath the stairs and gave the risers a light prod...

A word of explanation here: anyone who has ever done any serious salvaging knows there are two phases in this work: the meticulous board-by-board method described above — but at other times you have to go into a state of Creative Berserking. It's as if nothing can stop you, you're pushing down walls all about you amid great sturm und drang... and you're still standing there, crowbar in hand, covered with sweat and grinning through the bat guano and wood grime. Though sound and fury threaten to engulf you, you're indestructible.

Today I was in Berserker mode.

So... I figured well heck, I'll just give the stair risers another light tap with the crowbar, strike an Atlas pose and catch the thing...

Tap

When the stairs fell on my head, arms, and shoulders I remember thinking: this is what it's like playing rugby and getting gang-tackled in the middle of a scrum.

So much for Atlas. The stairs had fallen straight down on top Mr. Milquetoast, and I had simply folded up like an accordion, crumpled backward with the stairs pinning me to the floor. As I lay there under the down staircase, arms and legs poking out from either side, I had to start laughing... Zonk, that was one of the stupidest things you've ever done. I've heard of people falling down stairs — never stairs falling on people...

The experience left me with a Red Badge of Ignorance on my elbow where I had taken the fall. But it was nothing really, just "took the bark off," as the old-timers around here say. Less like a wound and more like a memento from the Giving House, a reminder from the old girl: "Don't take gravity too lightly."

Friday, Aug. 27, 1976 — We don't take much to graven images around here, and I've always thought it tacky to have concrete figurines in your lawn — but two have come to live with us down in the Holler, and they're symbolic.

The first is a cement rabbit so animated that we knew Hazel-rah from the first moment we saw him. Or maybe he's the Velveteen Rabbit. At any rate, he's Real.

The second was a gift from Jeanette. A fat, bald laughing Chinese good-luck god, Ho-Ti, I believe is his name. He's just the Zen reminder I need from time to time. Ho-Ti spends a lot of time sitting up in the woods laughing at his bald-headed counterpart.

Ho-Ti watches and laughs at his look-alike builder buddy.

11

In Which We Have a Son and are Illuminated

> *"Lo, children are a heritage of the Lord:*
> *and the fruit of the womb...*
> *As arrows are in the hand of a mighty man,*
> *So are children...*
> *Happy is the man that hath*
> *His quiver full..."*
> —*Psalm 127*

Tuesday, Sept. 28, 1976 — My son, Jon, has joined us on the outside this morning. It is late night now, and I'm back at the Blue Shed trying to internalize The Most Amazing Thing I've Ever Seen.

Words like stumbling messengers arrive babbling and incoherent. The moment is too great for them.

Who, having been there, can speak of the birth of his own children with detachment and clinical objectivity? For a woman it must be the ultimate physical experience; pain and ecstasy — and so much joy that writing about it sounds downright mawkish.

But still, the words must come out, just as Jon had to be out with it: a sight for all time.

It started early this morning at 5:30 with the water breaking. Excited, we jumped out of bed to see if this was really LizaJon's debut. And it was. A light pink show was found with the water.

We woke wide-eyed Selena with the good news. Then Maggie began having contractions ten minutes apart. After a quick Breakfast of Champions to fortify ourselves, we packed our bags and took off in the gray, misty pre-dawn for Crossnore Hospital, 30 minutes up the mountain. Crossnore is not only Maggie's hometown, but it was also where she was born, where Selena was born, and where she knows all the nurses, and they all call you "honey." It's the next best thing to having a baby at home. We had agreed this is where we wanted to have LizaJon, even though the old Doctor Fink (who incidentally delivered Maggie and Selena, and who is a second cousin of Maggie's) said she'd never heard of Lamaze or Rooming In... she did say she'd give it a trial run and if we did okay, then she'd see about it. She told me sternly that I would be the first man ever to be allowed in that hospital delivery room for the birth of his child.

We roared up the mountain at 70 mph while the contractions increased in tempo from five to three minutes apart. There we were, speeding along the narrow mountain road, laughing and talking, and when the contractions came, doing Lamaze breathing — and it was working.

We got to Nannie's by 8:30 and were welcomed by Aunt Cordelia and Nannie, who insisted on feeding us, so I wolfed down ham and eggs, but Maggie was ready to pass out, so when Nannie (88 years old and an veteran on baby matters) said we ought to get over to the hospital, I paid attention. In little Crossnore (pop. 250) going to the hospital means going about 100 yards across the valley, so we were there within the space of a minute.

By then, Maggie was having contractions every two minutes, and while nurse Aldie Johnson and Dr. Fink prepped Maggie, Dr. Fink grinned and said, "You're going to have that baby all by yourself," which is exactly what we intended to do.

Maggie was wheeled into the small, green tile delivery room, and I was ensconced in a mint green gown. We got Maggie on the delivery table and had a time insisting that Maggie not lie "flat of her back" in the traditional hospital pose. So we got her jacked up, and by that time, Dr. Fink found that Maggie was completely dilated. "Well, Margaret, it's up to you," she said, and, "You can start pushing any time now."

Maggie began, and I propped her up under her back and braced her while chattering at her and coaching her breathing. Soon Dr. Fink observed happily, "Well, it's got dark hair," and then 15 minutes later of pushing, we got him/her to the place where his/her head wouldn't slide back down the birth canal.

At last, Dr. Fink wondered aloud if Maggie shouldn't take a hormone pill to bring on a stronger labor, but I said I was not in favor of that, we wanted this to be as natural as possible, and that seemed to make Maggie redouble her efforts by shoving the baby into this world. But after two more unsuccessful pushes, I consented... but no sooner than had Maggie popped the thing into her mouth than she had a terrific push which shoved the top of the baby's head up and out. Dr. Fink performed a quick, small necessary episiotomy to the side of the previous one...

Then everything happened at once. While Dr. Fink said, "You can rest now," Maggie practically shouted, "How can I stop now?!"and kept shoving and shoving. In ten seconds the upper body of a lovely new person emerged and while only half out hollered its first cry... we were all working, pushing, catching and coaching.

I saw the umbilical cord and mistook it for a boy's parts, and had to chide myself to wait for the proof. The excitement and anticipation of what LizaJon would be was as unbearably delicious as Christmas morning.

Then he popped out, or I should say, slid out — like a baseball player in slow motion sliding into home plate where Dr. Fink, the Catcher, caught him. He was Out. But he was safe. And he was he.

When I saw his maleness, I hollered, "It's a boy. Look!" and Maggie could raise up enough to witness that small wholly holy perfect miracle, the birth of our son.

Amid tears and laughter, Jon was placed on his mom's tummy and I was instructed to hold his feet lest he tumble off. I thought this curious as it was not in our Lamaze training, but just then my son watered me down, but good. Dr. Fink stepped back expertly as a bright arc of urine christened the new father. Somehow I got the sneaky feeling this was not so unexpected to the old doctor, but at the time, it only produced more laughter and joy. "Well, that works," Dr. Fink quipped.

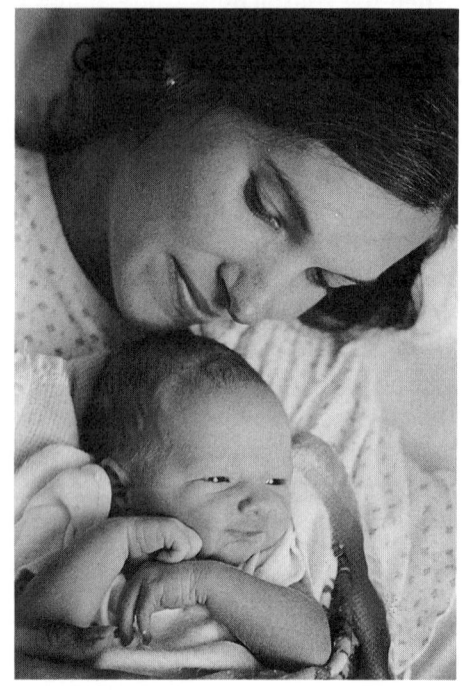

Words. No words. Only a baby's elf-blue eye and rockabye sweet baby Jon.

She cut and tied the cord and the placenta was delivered. Jon had some white vernex on him, but not much. They wrapped the little boy in a blanket and put him on the surgical table nearby, and I went over and hovered over Jon — for it was Jon at last — and did the strangest thing: I wanted to sing to him. But the only song that came into my delirious mind was a John Philip Sousa march. What the hey — why not a celebratory march at the birth of your son? So there I was, whistling and crooning *The Stars and Stripes Forever* to my newborn.

Jon winked open his left blue eye for his first sight of this world. He must have thought: "M'gawd, a bald dude scat-singing a Sousa march." No wonder he slammed his eye shut.

That didn't deter the triumphant father. I kept crooning softly and rubbing his back to get his lungs full and circulation going while Dr. Fink patched Maggie. It was a happy time in that merry little delivery room, with Maggie beaming across at me with those tear-sparkled silver love eyes, me humming Sousa, and Dr. Fink talking softly to the nurses and congratulations all around. We wheeled immediately across the hall into Maggie's room, and we had insisted that Jon come along and that I be allowed in as well; this "rooming in" was all very new for this mountain hospital. But we had been a hit with Lamaze, and Dr. Fink said she was so impressed with the method that she would allow other trained couples to try it. We felt like true groundbreakers — as well as babymakers.

I swear but that Jon smiled in the low light of the hospital room. The first thing I did was take their picture. We spent the day with each other marveling, phoning friends and family, admitting family members to gape at the newest news.

But it was Tuesday, and so I had to blast back down the mountain and help get the paper out. At 12:30 a.m. we finished up, and I charged back up the mountain to see the wonder of it all.

Wednesday morning, I found that Maggie had not slept a wink since her water had broken. She was in a state of grace, floating in a transcendental joy that she didn't want to surrender to sleep.

She said: "If this was Heaven
  then there would be
    no pain.
But this is Earth,
so it is
  rapture."

Saturday, Oct. 2, 1976 — I have returned to Crossnore with a batch of birth announcements I composed and had printed. Sitting on the bed, we addressed the cards and chattered happily to each other and Jon. Selena has been a marvel too. She announced today, "I'm gonna hold that baby until I get sick and tired of it."

Jon's first trip to Hogwild. Ruby guards while Jeanette gives the girls and dolls a ride on Tony.

Sunday, Oct. 3, 1976 — We brought little Jon home today. The sun is out. The little nursery we fixed up for him in a back room is small but delightful with a high crib and wallpaper that reminds me of all *The Wind in the Willows* critters at a country fair. When I hold him in my arms, I point to the animals and recite their names, there's outrageous Mr. Toad, stout-hearted Mr. Water Rat, and heroic Mr. Mole. Jon will know more of these characters later.

Son-rocking at midnight in an old yellow wicker chair, my reeling senses are barraged by unedited overlapping phrases and quotes from the week: For unto us a child is born/ hail to thee, little man; barefoot boy with cheeks of tan/ all babies look alike — like Winston Churchill/ Uncle Ben says it's a 'possum/ he's got her nose and my eyes.

Winking smiling waving crying my son speaks perfect body language.

A leaf, a stone, a baby's elf-blue eye
and rockabye sweet baby Jon.

Sunday, Nov. 1, 1976 — This year, by gum, we're going to be ready for the winter and wood heat. Having two kids in the Blue Shed has mobilized me into preparing adequately. If I'm going to heat with wood (and in the Blue Shed that's all she wrote) then I've got to be serious. Last winter we stacked wood on the porch and it was better than nothing — but an honest-to-goodness woodshed would be ideal.

So, for the last two months I've been banging away on this critter — a project designed to keep me building close to home, since Maggie hasn't ranged very far from the Blue Shed with Jonathan Rush on her hands. He demands and deserves our attention, a fine lad, we're so proud of his even Libran temperament. A "good baby" by all standards, not colicky like Selena was as a poor babe. I reckon she was miserable as an infant — I know we were; it seemed to me she cried all the time.

I should call the *Guinness Book of World Records* about this woodshed/toolshed/glorified clubhouse of mine. I have been keeping track of expenses and by my calculations, I built it for $7.50 — and that was just for nails! The salvaged wood came from several sources: Father Austin's office remodeling two years ago, the exterior siding and the door come from the Giving House, the windows were freebies, and the roof — ah, here's the coup de grace — it is aluminum sheeting which I got from the newspaper pressroom. They're the plates actually used to print the offset newspaper. I simply turned the ink side down and nailed them in place... I bet I'm the first builder ever to get printer's ink on his hands while laying down roofing. And it feels sort of strange to be in the woodshed and look up at the ceiling and be able to read last week's newspaper.

I built the woodshed on a concrete slab that was left over in the side yard from some previous owner with a trailer. When it was done, I discovered to my chagrin that I had built one corner narrower by a whole foot. Duh. So the little woodshed has a decidedly whackeyjawed look about it. Oh well, each outhouse is another dry run, and I recall a

132 *Hogwild*

quote I need to put down but can't attribute: "There are no mistakes — only lessons."

Wednesday, Nov. 3, 1976 — For the last two weeks Hogwild has been the site of a magical event: a watercolor workshop led by our artist friend Tom Cowan.

Bill and Jeanette served as co-hosts, cook, and maitre d', stove-stoker, and house musician for the 27 artists from all over the country who stayed at the Big House.

We regard our friend Tom Cowan as the best North Carolina watercolorist, but he has not received the recognition we think he deserves. Like many artists, his strength doesn't lie in marketing and promoting himself. I think Jeanette and Bill came up with this workshop idea in order to help him out. Tom has been living out here from time to time — in a converted schoolbus parked up in the woods near a mountain I call Sundowner, or for a time in the spacious hayloft of the Byers' horse barn.

Swaddled in wrappings, artists take advantage of the clear morning air.

Tom, a native McDowell and Rutherford Countian, went up to Chadd's Ford last year and had a chance meeting with Andrew Wyeth, his idol and mentor. The meeting forever changed Cowan's life. "Paint what you know best — your home," Wyeth reportedly told the young artist.

Tom took Wyeth's advice, returning to the mountains and to Hogwild where he set up a summer studio in the barn loft at B&J's. There, he started seriously cranking out watercolors as fast as he could go. And they are so innately Hogwild. It's clear that he has found a rustic inspiration here.

So it was only natural that a workshop be held in the Southern Gothic splendor of the many-cupola'd Big House — a structure I've

Tom Cowan instructs neighbor Billy Whitesides, a budding artist.

always said reminded me of a Mississippi River boat (minus the paddle wheel) marooned up on a hill.

After advertising in prestigious national art journals, the workshop drew folks from 10 states, and we all enjoyed the paradoxical sight of Cadillac Sevilles, Jaguar XJ-6s and Mercedes clogging Hogwild's narrow dirt driveway.

The students themselves included a coughdrop factory worker from Pennsylvania, a retired Naval commander of the nuclear submarine base at New London, a Green Hill schoolboy (neighbor Billy Whitesides who has been excused from classes to pursue this interest), an interior designer, a hairdresser, a pastel artist from Disney World, and several women like Pat Viles of Conover, N.C., who Saturday, when it was all over but the hugging, said: "At 4 o'clock today I go back to being a mother and a housewife."

Jeanette cooking in her big country kitchen.

The artists experienced a glimpse into a life of wood heat, down-home country music, home-cooking, no television — just two weeks of thoughtful watercolor painting. A few of them did some lifestyle re-thinking. "Now this is the way to live," concluded one svelte big-city Northerner.

Tom would give instruction on technique in the barn loft studio in the mornings, and then in the afternoon the artists fanned out across Hogwild to paint and sketch on their own while Tom made the rounds, giving each one individual attention.

They painted the Rocky Broad, a teepee in the meadow, the barns, the Big House, all made more amazing by the clarity of the October chill, the tang of woodsmoke, and Hogwild's raucous autumn colors which were at their best this week — as if the Land knew it was modeling and wanted to show off.

Christmas, 1976 — We are truly woodbutchers. For Christmas, Maggie has given me a big red wheelbarrow to tote wood from the new woodshed to the house. With four wood stoves going, we are able to keep the Blue Shed right toasty in spite of old leaky windows, no subflooring or insulation. When I'm at work, Maggie keeps the fire going in just the living room-kitchen combination, where the Franklin stove keeps this "snuggery" livable.

In Selena's room we have a Fisher "Baby Bear," in the front living room we use only on special occasions, we have a wonderful old stove we call "Vesuvius" and in my office where I do most of my writing on an ancient pre-World War II Underwood, I have a decrepit little stove I got for nothing at the demolition of an old house downtown in Rutherfordton. The Blue Shed, with its two tall chimneys, has flue openings in every room except Jon's where we've been forced to resort to baseboard electric — a grudging concession to necessity and modernity.

Obviously, the first word that Jon has had to learn is: "Hot!" and as we point to the stoves and jerk our hand away in mock alarm and we say it with such wide-eyed gusto that at three months the little elf has learned "not to touch!"

He is full of smiles and chortles for all of us, this alert happy little character. Selena has emerged from her self-absorbed only-child state, and has made herself precious to us more than ever now in her new role of Big Sister. It is a role she has embraced with gusto, and I'm very proud of her acceptance, and indeed, of her ownership in this experience.

We are at Nannie's big house high in the mountains where it almost always snows for Christmas. No less than 25 kith and kin are here.

In Which We Have a Son   135

This afternoon a slew of us took the traditional post-dinner trek across the mountain over to the "Uncle Robey Place" when a light misting of snow began and then started falling heavier. The mountains' tree-lined gray ridges stood like a steel etching against a sky falling in silent slow motion. We stood in a whitening meadow, listened to the hush of the evening snowfall, throwing our heads back to catch snowflakes in our wide open mouths.

Selena and Linda Whitesides play in the snow on Slickey Mountain; the view shows her playhouse, the Soccer Field, and Bill and Jeanette's place beyond.

12

Wherein the King of Siam Builds a Chimney

> *"If I am not for myself*
> *who will be for me?*
> *If I am only for myself*
> *what am I?*
> *And if not now... when?"*
> — Hillel,
> The Ethics of our Fathers

> *"I yam what I yam."*
> — Popeye the Sailor Man

Saturday, Jan. 1, 1977 — Ok, here we go. A new year and the first day of another Spring Offensive. Over Christmas we have decided to take the leap: we're aiming to be in the cabin by next Christmas.

This is outrageous, but if I don't give myself a goal then this thing could drag on indefinitely. It's already been three years since we started dreaming, and two and a half years since construction started.

So we have categorized the remaining work in order to "divide and conquer." Low-cost, single-handed activities: rockwork — all I need is mortar, flue-liners, freebie junk rock for fill, more firebrick and trowels; flooring — need to get boards planed and buy nails; framing — entrance way, bathroom and hall, just need 2x4s.

Moderate cost or two-man activities: deck — have three long beams, but need pillars of some type, as well as decking itself; stairs — Carlton Mason says he'll do it for cost of lumber; insulating floors — would like to have help on that nasty project.

Expensive stuff: plumbing, wiring and windows.

After examining the list, it seems the thing to do next is to frame the interior and start on the hearth and chimney by myself as it seems that

Kurt has disappeared for good. The last couple of times out here, he'd say, "My head's just not right to lay rock today." I reckon what he meant was that he really wanted to go write his dissertation on Soren Kierkegaard. I understand Kurt's moved to Black Mountain and is working on his Ph.D. Regardless, I owe this gentle philosopher/rockmason a debt of gratitude. He may have left me with a chest-high chimney, but he also inadvertently taught me how to lay rock.

Sunday, Feb. 6, 1977 — This week I have been framing the gables over Selena's room and the bathroom, the hall on the north wing, and the entrance way to the cabin.

I sense some sort of a sub-religion, or at the very least a guiding philosophy emerging during this journey. I am metamorphosing from a naive boy who thought he could build a cabin in 90 days to a young man burdened with the awesome weight of his commitment. And something else I am learning to embrace: I am alone. Ultimately this cabin a'bornin' is my burden. Maggie is there to coach me with the breathing, to support my back, and cheer me on. But this cabin is my baby. Right now I feel like it's stuck in the birth canal. I'm in prolonged labor. And this year is going to be a major contractor's contraction.

I am alone. One man versus height and weight and gravity with no blueprints and no team of carpenters to help me heft the other ends of things, measure places I can't reach, hand me tools I've forgotten or dropped when I'm perched on ladders or dangling off scaffolding or hanging off log walls; no one to jaw with me and keep me going; no one to tell me what to do next. Hence I repeat to myself: "What next, Spiderman?"

I am truly single-handed. A carpenter with but one hand.

I'm working at heights and over spans and with weights which defy the rules of modern commercial carpentry. Sometimes my own self-doubts and realization of my own abysmal ignorance of building are my greatest obstacles. Who would put up with a contractor/builder who has taken three years to build a house? I'm not sure that even I can put up with myself much longer.

When I say I'm working without plans, I mean formal plans. We've got gobs of sketches, doodles and diagrams done on the backs of restaurant checks, city council notes, and paper napkins which constitute the body of our "plans" as such. But they're just broad brush strokes, and the mystery of translating ink on paper to the reality of wood and nails, rock and mortar, glass and plaster is left up to me.

I have learned to cope by using a meditative process of "imaging," or even better, en/Visioning. That is, I stand before the project at hand in total silence, staring at the space where this wall or window or chimney must go, and I get visions in my mind's eye of how it will go up, piece by piece,

When I'm in this mental state, mode or brainwave pattern of the super creative builder, I find that I am having a certain builder's high, or a woodbutcher's buzz. No wonder building is addictive. Sometimes I wonder if that's part of the reason I seem to take so much time for each project.

Lessee, I'll try this, and then that, and then...well, don't worry about that third or fourth step, just do it. And I come out of my Architect mode and go into Carpenter mode, but all the while the Visioner is watching over my shoulder, making suggestions and refinements along with way. Ha, that worked. Now, on to the next step. Ah! this is so logical.

It's like I'm right-brained, mostly intuitive and not very mathematical or rational. I wasn't much of a scholar, and always regarded myself as "a bear of very little brain." But when I'm in this mental configuration — imaging and creating from the ether of visions — it's as if I'm using a whole bunch of gray matter on the logical side that has never before plugged in. It's like driving down the road on three cylinders and the other three suddenly kick in. And the feeling is positively electric, as if the synapses are literally clicking like circuits firing; the mental power surge is almost frightening.

Into the equation I must always factor in my singularity and how I will manhandle weight, height, reach, and span. Have you ever watched a spider on a ledge tentatively extending one leg or another over the chasm, and then sending out hopeful webbings? How do you accurately measure across 12 feet of space when you're 14 feet up in the air? I become the Spiderman sending out my measuring tape like the strands of webbing over and over again until it reaches its destination and takes hold.

Saturday, Feb. 12, 1977 — I know I don't have time to do this, but I couldn't miss the opportunity: I'm playing the part of the King of Siam in the local community theater production of *the King and I*, early in April.

I can't pass it up because it's the ultimate stage role for the likes of me. I mean — how many chances does a bald dude get to star in a major production, strut around the stage like a big spoiled brat, holler at people and have them jump? Hey, I'm beginning to like this King business.

Wherein the King of Siam Builds a Chimney 139

Maggie with baby Jon and Ruby — and in the background a chimney that has started calling my name.

We've been in rehearsals about a week now, and already Maggie is telling me I'm becoming impossible, so I reckon that means I'm starting to get into my character, that of an Oriental irascible curmudgeon, lovable yet unlovable.

140   *Hogwild*

At the outset of the renewed campaign on the chimney, it looked like this. Feb. 20 — at last, I'm rocking on again.

Sunday, Feb. 20, 1977 — The King of Siam stalks around his log cabin kingdom learning his lines, shouting royal proclamations to the trees and ordering the rocks about.

Some of the stones are obeying. I noticed one rock that seemed to want to go... here in the chimney. And just like that, after a year's hiatus, I am rocking the chimney in earnest, having decided at last that if it's to be done, I must do it myself.

If it's good for the cabin then it's good for Siam. The King is losing weight, and this beats pumping iron. If I'm going to walk around on stage bare-chested in front of half the county it might look a little strange for the Lord of Light to have a beer belly hanging out.

Wednesday, March 2, 1977 — We have taken a loan out on the Blue Shed and now have enough shekels to shake this cabin project loose. First, we ordered $1,500 worth of Andersen double-glazed, casement windows. Next, I found old Leroy Sims, a local plumber, and he came out and plumbed the place for a couple of hundred bucks. Likewise next week, a local electrician is going to wire the entire cabin for $400. Meanwhile I continue rocking along and pretty soon I'll need to rent a second round of scaffolding.

Friday, March 18, 1977— After three separate whole days of rocking I'm up over six feet now. I'm filling in the cavity between the flue liners and

We've just put in the new windows and hope they're worth the money.

the exterior rock with junk rock I'm picking up along the roadside — and when I find a nice big flat slab of granite, I put it aside for the hearth.

Today also, the windows were delivered, and we started putting them in. The biggest one of all, which is roughly 5'x9' and about 10 feet off the ground, was hauled into place in the kitchen with ropes by Maggie, Jeanette, and myself.

Electrician Terry Wilkie came this week and wired the place in one morning — and wiring a log cabin is no simple chore. We had to figure

142 *Hogwild*

The rockmason turns King of Siam briefly as my Yul Brynner haircut pays off. My leading lady is Ursula Hunt of Rutherfordton. (photo by Gerry Gerlach, courtesy of Rutherford County Arts Council)

out together how to run power over, under or through logs so the wires would be concealed. The solution was to go under the floor and have mostly floor outlets in the living room which is within the main core of the logs. Then he wired the two "modern" wings conventionally.

Sunday, April 3, 1977 — "The King is Dead. Long Live the King!" After three stunning performances playing to packed houses in the Cool Springs Junior High auditorium in Forest City, I have laid to rest King Mongkut, the Lord of Light of Siam, circa 1862.

The backwoods Yul Brynner perceives a nuance in the title of the

King and I, as if the King is another facet of my self (as indeed he is) and the play is titled by the actor playing the King.

I can see now why drama folk get so completely hooked on show business. The performance demands a total effort from you, and in return, when you're on stage (and there really is something to "taking" the stage) you get a soaring sensation. For instance, I'm no great singer, but walking out on the stage for one of the King's solos with the music of a full orchestra lifting, supporting, and encouraging me, I was filled me with a sense of responsibility and power.

Vignettes from the play as they quickly recede: right before the opening curtain in the gloom of the darkened stage, three Siamese dancing girls and me all doing deep knee-bends, sit-ups and push-ups to get psyched up for the show... once having to improvise lines when someone on stage skipped two pages ahead in the script, the panic in her eyes when she realized what she had done, and then relief and triumph when we steered the scene safely to its proper conclusion without ever breaking character or letting the audience know of our mental gymnastics... finally the big death scene of the last performance when I was so totally in character and died so thoroughly that I just lay there and forgot to get up for the curtain call — then when I realized the curtain was opening, Lazarus had nothing on me. I vaulted from by deathbed and raced the opening curtain to the stage right wing — and ran smack-dab head-on into one of the palace walls.

The dark bronze makeup has lingered all week. Sunday night after the last performance I took a final bath in the Blue Shed and the "race" that I had become for a week was e/raced, and all that tan was committed to memories and bathtub rings...

But I know the cure for the post-play blues. It's head-first back to cameras, typewriters and my real Hogwild kingdom in Sourwood Holler where my palace of logs awaits the return of the king.

13

Rocking On

> *"Life always gets harder toward the summit."*
> — *Nietzsche*

> *"L'audace... toujours l'audace."* *
> — *George Jacques Danton*

Wednesday, April 20, 1977 — My royal dais is now my scaffolding extending up worshipfully in front of the rock chimney. I have drawn several designs for the chimney and I can see that the hardest part is going to be to make the angles match as the chimney turns in and angles upward at about the top log. How on earth am I going to match the angles? Is there a rockmason's tool I don't know about? So far all my chimney designs look like soft drink bottles.

Wednesday, April 27, 1977 — After a weekend of modern dance workshops with New York-based dancer Dan Wagoner, we had him out to the cabin for a picnic down by Silver Creek. It made me see how truly rustic and primitive the place is, seeing it through the eyes of someone from the Big Apple. But instead, the world-class dancer from the Big City could only say nice things about the clarity of the air and water and of our purpose. We were moved.

Friday, April 29, 1977 — Framed in gables over bathroom and Selena's room. Installed windows, put in plywood, insulated with fiberglass, nailed up tarpaper and shingled the exterior with cedar shakes like the roof. It looks terrific.

Saturday, April 30, 1977 — Thunder at Sourwood Holler. Throaty ominous stuff makes Ruby the Weather Dog pant-pant her hot, nervous

---

* *"Audacity — always the audacity."*

breath on my bare leg, slightly crazy red eyes implore me to make the loathsome shotguns leave her alone.

Spring's first frogstrangler complete with incessant rolling thunder booms in on sheetfeet of sound and rain across my valley.

All those new leaves will now clap open like popcorn. Now they are wagging at being patted reassuringly by sky strokes.

Neighbors Bill and Jeanette have been caught in the downpour while investigating a break in the waterline up Spring Creek. Now they are seen running through the Soccer Field like two kids playing kick-the-rain in the puddles. Their two horses, Tony and Elfie, who have been grazing in that meadow, go stand in the eaves of Selena's Playhouse.

I watch the rain and feel very much at home.

Saturday, May 7, 1977 — I've heard sculptors say that the rock will tell you what it wants to be, that if you'll listen close enough the form within will dictate the carving.

Now I've learned something of a similar law about laying rock: that you let the rocks tell you where they want to go and what they want to

Chimney drawings, real and fanciful.

Spring comes to Sourwood Holler, as father and son cut up.

do. But you've got to trust the rock implicitly.

Here's how it works: I stand looking first at the space I must fill. Then I walk around the rockpile like a dowser in something of a meditative trance until one rock will...well, for lack of a better term — glow. The right rock will actually seem to be emitting some sort of low-level radiation as if signaling me.

Maybe it's Zen and the Art of Rockmasonry, but you'd have to be made out of stone yourself to miss that energy. The first time this happened, it was slightly scary. But then I learned to trust the rocks and await their signals. As a result I'm now 10 feet up and going strong. Actually the hardest thing about all this is hefting the huge flue-liners in place. The terra-cotta monsters are big, ungainly, and must weigh 80 pounds. I have to wrestle them up the scaffolding and then into place in the center of the chimney.

Saturday, May 21, 1977 — It has taken me several days to figure out how to install the wooden sliding double doors made by Peachtree that face out onto the creekside deck. Because I needed the 6-foot, 9-inch height, I was forced to drop the sill down to the foundation, and chisel out half of an overhead log. When it was done and in place, I was just thrilled by the results.

Saturday, June 4, 1977 — I reached the top log today. That's 12-feet, six inches up. Now it's time to swing in on the angle before going up again.

I have come up with a primitive rockmason's angling device. Remember how worried I was about determining my angles and duplicating them evenly on each side of the chimney? I've devised a simple device that looks like a "T" with a moveable crossbar. It's so simple that it's almost sinful, made out of two 1x2s. I stand back from the chimney, hold this device up to the chimney, and literally en/vision the rocks' angle, then duplicate it by matching it with the T's cross. I nail it in place.

Then, I pick out my rock and carry this angle tool with me up the scaffolding and simply use it as a guide when laying the rock. I'm sure real rockmasons have an official tool for this function, but out here we're talking about the mothers of invention.

Sunlight highlights the lines of the face-laid rocks.

Thursday, June 9, 1977 — Framed in the gable over the chimney so I can rock on up. And in an act of acrobatic courage, I climbed out onto the top log and knocked out the original round, crooked kingpost which supports the center of the roof's gable on the uphill side, (and the roof didn't fall down as I had fantasized) and replaced it with a pine 4x4 salvaged from Dr. Hyde's cabin years ago.

Friday, June 10, 1977 — I have been so focused on the chimney I didn't even remember my birthday Wednesday — but Maggie did, catching me totally off guard with a surprise lawn party at the Blue Shed when I returned from the cabin that afternoon.

Maggie had sent out the word that what Jock really wanted for his birthday was rocks. So there was our front yard, littered with stones of all sizes and shapes. My buddy and fellow reporter Joy Hoagland brought me a 50-pound hunk of granite tied up in the demure pink ribbon.

And then there were Hogwild neighbors Bill and Nancy McCullough who maintained they thought they were supposed to bring a cock, not a rock, as they appeared on our doorstep with a lightning-eyed rooster with fiery feathers and bristling for a scrap.

Said bird promptly ran a'fowl and took to the neighbor's pine tree and I spent much of my birthday party climbing same and picking pine needles out of my hair, such as it is. After thanking Bill and Nancy, and convincing them of their mistake, we bagged the bird and deposited him in the trunk of their car.

So what should my wonderous ears perceive the next morning when at dawn's early light there broke upon the downy dusky mantle of morn a raucous Cocka-doodle-doo! coming from the side yard. Hmmm, I thought vaguely as I wandered about the Blue Shed's dew-piddled lawn searching for the morning paper. Hmmmm, what a pleasant pastoral sound is that. Reminds me of Chanticleer...

Yoiks and oddsbodkins — it dawns on me, that's my rooster. My birthday rooster perchen in yon woodshed waking all my goode neighbors. Avast ye squabs, ye dastardly deed-doers. How cometh it to pass that yon crowing cock hath extricated self from cartrunken and tis belocken in mine olde woodshedde? 'Tis passing strange methinks.

Three more trumpet blasts from that rooster and my mind's made up. Donning my Rooster Gloves, I grab Brewster the Rooster by his dangerous heels and heft him into Obie the Rooster Bus (which also doubles as a duck bus, dog bus, photo bus, log bus, rock and cow manure-totin' bus). The dang rascal crowed all the way to my log cabin.

Brewster the Rooster has gone Hogwild where he can crow all the live long day for all I care. Yes, I feed him, and he sleeps under the cabin porch. No, I don't have a dab of pity when Maggie suggests, "Don't you think he'll get lonely?"

To paraphrase someone famous: Frankly, Scarlet, I don't give a hoot.

Saturday, June 11, 1977 — While framing in the gable over the chimney

I banged one of the top rocks with a piece of plywood and the rock toppled, struck me on the shoulder, glanced off and fell with a great crash to the rockpile below. I cussed like a sailor. I'll be okay. I think I was more stung by my own clumsiness than I was by the wound itself.

In the late afternoon a summer trashmover rolled in. The squall line blew in pitch-black from the west like some aerial tidal wave. I covered the chimney with the tarp and concluded it was wise to get off that lightning rod-like scaffolding.

It is three years to the day since having begun this journey in wood, and now in rock. That seems much more significant to me than the fact that I've turned 32. How can three be more than 32? But it is.

While I'm rocking on, Maggie has undertaken the insulation of the walls because she can reach it and the work doesn't involve heavy lifting.

We've yet to run water, lay floors, finish top gable windows, rock the hearth, finish the inside walls, do the entire kitchen cabinets, install lights, build a stairway, construct loft closets, finish the deck, and I'm sure there are 10,000 other things I've forgotten.

That's me, looking like a man from Zircon, according to my daughter.

Thursday, June 16, 1977 — Down in Forest City the Kiwanis Club has been collecting rocks from all over to build a memorial chimney on the town square to commemorate the town's original name of Burnt Chimney. Last week they finished the monument and they had left over a slew of rocks that folks had toted in from vacations all over the nation.

Gene Blanton of the Kiwanis Club knew that each one of those rocks was in some way significant to its contributor, so Gene called me and said he wanted those rocks to go to another historic chimney and that I could come get them. I was only too happy to get all those pet rocks for my pet chimney, which is in itself a monument to one man's audacity.

My budding artist daughter presented me with a drawing of the budding rockmason at work. I think it captures the essence of the character rather succinctly, particularly the fact that she has depicted me as one of Erich Von Daniken's outer space crazies.

Well, if the shoe fits....

Thursday, June 23, 1977 — I am charging now. Last weekend I did the angles and my homemade angle device worked like I had hoped it would. That phase went quickly because I didn't have to work against gravity; the rocks just lay there, as opposed to the ones I'm laying now, which must stand on their own, pretty much free of mortar. I mean, I'm using mortar but it's deeply recessed and I've learned that if a rock is going to stay put, then it's got to fit and feel right for all time. The best way to lay rock is to pretend you're doing it drywall, that is, without the crutch of mortar. That's the way the Incas and the Egyptians and the old masters of pioneer America did it, drywall. Remember the first rule of great rockwork: there ain't no cement in the Pyramids.

As of the end of the workday I am three rounds of scaffolding up in the air and ready to cut into the actual roofline so that I can go higher, and on up above the angled eaves. I might even make it by my target date of July 1!

Sunday, June 26, 1977 — I am at home at the Blue Shed, totally exhausted. It is dark now and I can barely write. But I've got to get this down. I cut the roofline with the circular saw on Friday, and then the monolithic thing rose as if by its own volition over the roofline.

Saturday I had to add a fourth round of scaffolding, and now that puts me about 25 feet in the air up over the rockpile, increasing the angle of ascent to the point that yesterday afternoon while grappling with a humongous flue liner, I about fell to my ruin. I had gotten the critter to the third level of scaffolding and was heaving it up to the fourth and last... when I realized its weight was bowling me over backwards.

Everything goes into slow motion in my memory banks at this point. All I can remember is that I knew it was a choice of dropping the flue liner (about 12 bucks down the tube and a trip 16 miles back to town for another one) or trying to save myself with some sort of weird maneuver and in the end probably getting killed trying to save a freaking flue liner. I concluded in slow motion that it probably wasn't worth the risk, and as we (the flue liner and I) were falling, I heaved it away from me well out over the rockpile, and at the last moment grabbed the closest iron bar of the scaffolding. The flue liner hit the rockpile with a dramatic explosion of terra-cotta on granite.

A red warning light was flashing. It was Ratcliffe's Rule of Building — Number 47, taught to me by my old carpenter buddy three years ago when we were charging along on the roof late one afternoon and Charlie started banging his fingers with the hammer and almost fell off the roof. Looking at me, he said levelly, "When you start hurting yourself, it's time to quit." And so saying, he climbed down off the roof and hung up his hammer for the day.

I decided to heed the alarm, thus ending a day's work. At least the only thing broken was the flue liner, and I returned to the Blue Shed shaken but intact.

Tonight, looking for spiritual refreshment I have turned to the *Tao Te Ching*, and have been warmed by Lao Tsu's admonition reaching from sixth century B.C. China across the ages to a remote Appalachian mountain holler where a sun-burnt, bald-headed, bedraggled specter in cut-off khakis and brogans hears the wisdom:

"People usually fail when they are on the verge of success.
So give as much care to the end as to the beginning;
Then there will be no failure."

Friday, July 1, 1977 — I have missed my deadline but don't care, because I'm doing so well. I've worked through horrific heat and monsoonal downpours these last two days. My fingers have become raw on the pads from handling the skin-eating mortar and abrasive granite. But now I'm too excited, too near the end to care about a little pain.

Sunday, July 3, 1977 — Today when I laid the Last Rock I didn't even realize it. I was concentrating so hard on not falling out of the sky, balancing there 25 feet over the rockpile, one foot on the ladder, the other on the top of roof's ridgeline, juggling rock and mortar, trying not to knock down what I'd just set, and getting the top relatively level... that when the last rock was in place, only then did it dawn on me: that was it.

Selena was playing in the sandpile below and Maggie was watching, taking pictures as we cheered.

We couldn't wait to take down the scaffolding and see the lovely thing all by its self. Took it all down in a hurry, steel-brushed the excess mortar in the crevices, swept the spilled mortar off the roof, took down

152   *Hogwild*

Is it actually done? I feel just the way I look: completely wasted.

the ladder and stood back and just plain gawked. I couldn't believe it.

Forsaking the usual bottle of champagne, I kissed the chimney, hugged it — well, why not? — and sloshed a beer on its rocky flanks,

Our best working drawing, traced from a photo and filled in.

then gave my sunburt head a similar blessing.

Then for the longest time, I just sat in the shade in the driveway staring at the thing, trying to separate my fantasies from the reality of its form, trying to believe my eyes.

It was done. I had done it. My first impulse then was to record it on film, and we burnt up two rolls of film just to make sure it was real — at least on film. I wanted proof that it wasn't some figment of my imagination. I built a 25-foot-tall rock chimney. There! See? It sounds like an outrageous lie!

But yes, as the afternoon wore on, I began to accept the fullness of the new thing itself, slightly flared at the top like an old musketeer's blunderbuss, thrusting out of the earth more like a purposeful granite cliff rock formation than something man-made.

And then a new awareness began to steal over me: I hadn't built it. It has always been there. Always. My cabin has never been without its rocky companion. The cabin like a boxing ref holds up Rocky's triumphant paw over its roofline.

And I in the amphitheater of Sourwood Holler am on my feet wildly applauding the granite-jawed slugger.

Thursday, July 7, 1977 — Another first. These last two days I've started the first interior finish wall work. I'm grateful to be out of the weather, for July has turned terrifically hot and we've hit a dry spell.

The wood I'm using is the beautiful, old heart pine tongue and groove from the Giving House which Maggie calls honeywood because of its rich golden color. Working in the west bathroom wall, I'm laying the wood on an angle and it feels very adventurous and slightly zoomy.

Friday night we all "camped" at the cabin, and it was little Jon's first night here. I reckon it was a pretty crazy thing to do, for the skeeters about carried us off — yet sitting around a small picnic campfire, singing songs with Selena and Hannah, Ruby the Guardian Lion just at the edge of the quilt, and Jon drooling at the fire and waving his arms at the music... well, what's a few bugs at a magic moment like that?

On a cool summer's eve, we share the campfire's mystery.

14

Hearth and Homing In

"Nothing great was ever accomplished without enthusiasm."
— *Ralph Waldo Emerson*

*"Give me a fulcrum on which to plant my lever,
and I will move the earth."*
— *Archimedes*

Thursday, July 14, 1977 — It actually feels like we can get the house done in early winter. Wouldn't it be magnificent if we could make it by Christmas '77?

Now that the chimney is done, I'm filled with a new awareness of the possibilities. With 20/20 hindsight, I realize that the uncompleted chimney had stood as an obstacle. The task loomed so huge I couldn't see past it. But now, if I can just keep up my energy and keep plugging away....

The last two days out here I did more interior wall angle work in Selena's room, and about finished the bathroom walls.

Friday, July 15, 1977 — It's been hot as an oven these last two weeks. I've been walling Selena's room and sweating a six-pack a day, which is considerably more than my intake.

Today while mowing around the cabin I ran over an underground yellowjackets' nest and got blasted on the left hand before I could leap out of harm's way. The swollen left hand made working difficult the rest of the day.

Thursday, July 21, 1977 — What was that noise? At first I didn't even recognize the sound of rain. It's been three dry weeks. At first I thought it was big fat grapes plummeting out of an overhead vine: pat, pat, they plopped solo in the fine dust.

I turned my face skywards to the steel gray ice cream cone piling up over Hogwild — and received a solitary July baptism right in the eye. I laughed with delight. The emaciated earth seemed to gasp in relief.

The tempo of the patter increased as the lazy raindrops multiplied and grew finer. Up on the hill above the cabin the pines moaned a thankful sigh. The stuff began coming down in sheets, and the air became zingy and buoyant as the dust and gunk and heat were magically swept away.

There have been 19 straight days of temperatures over 90 degrees and no rain. If I'd been up on that scaffolding still working on the chimney, this journal would have to be completed by a ghost writer, literally.

I was so thankful I cheered unabashedly. You can do that when you live way out in Boondonia. Blessed water. Let it rain.

Monday, Aug. 1, 1977 — Remember Carlton Mason — the dude who said he'd do the circular stairs? Well, late this afternoon he came by with a load of beautiful thick and broad white pine.

I skipped City Council meeting to watch and help where I could. It's terrifically exciting to see it go up. We used one of the massive hand-hewn 4x6s from the Giving House as the upright around which to build the stairs. Carlton really proved what a master builder he is, mortising and notching the stair treads into the big upright. We stopped at 11 p.m. after getting seven treads in, and five more to go. I have no idea when he's coming back, but that's okay, I've got fifty-eleven other things I need to be doing.

Thursday, Aug. 11, 1977 — We've made a commitment to solar energy and appropriate energy use. This cabin will be heated almost totally by wood and sun. I want to try something they're calling "passive solar," which amounts to south and east windows, all thermal-paned (thermopane), deep overhangs, and no windows on the cold north side where the sun never shines.

Another thing I've learned about passive solar is that if a house is well insulated, and it is heated, then it needs something to actually hold the warmth. Our solar zealot friend, Rick Pratt, insists lots of interior rock in a hearth as well as the chinking itself will constitute what he calls "thermal mass" which will act as a holding agent.

We've decided to have high overhead windows in the main room gables, and Rutherford County Glass Company is custom-making thermal-paned windows to my specifications after I framed the openings and took careful measurements. Three windows for $115. Mr. Lester's prices are excellent and his outfit does good work.

Friday, Aug. 12, 1977 — The completion of the chimney has filled me with such confidence that I have now embarked on a truly Hogwild journey in rock: building a raised hearth and arched fireplace. Hello, of course I don't know how to build an arch, but I bet I can do it.

I only know after doodling some designs for a conventional level fireplace top I didn't like them. So I've got to figure out how to lay an arch. I've been looking hard at the magnificent arched bridges along the Blue Ridge Parkway — now there's some mastery in rock.

Today I started the simple part, which is just building the hearth itself, which will be raised one foot above the floor. This whole project is made possible by there being enough rock left over from the chimney for me to build this whopper inside. But the mountain granite is so irregular that I've elected to use local granite which I've just found prospecting along road cuts. And philosophically it's satisfying to use the free rocks of Hogwild in the hearth.

Note: I am committed now more than ever to finishing by Christmas. I've promised myself we're moving by then — no matter what.

Sunday, Aug. 21, 1977 — How about that? I finished the hearth today with the flat native granite and it looks fabulous: three feet deep, seven feet long and one foot off the floor. Now I'm ready to start going up with vertical rock. This will be like the chimney work but much more painstaking.

Saturday, Aug. 27, 1977 — Culture Shock. We are sitting on a deck at the North Myrtle Beach Hilton like two backwoods yahoos come to the Big Apple. I can't keep my mouth shut; this can't be real.

But for our 10th wedding anniversary, we were awarded a "free weekend vacation" by a local wood stove company. Well, it turns out that the whole thing is a trick and a treat from our friend Russell Scruggs who saw that we were going loonytune working on the cabin all the time — and the only way we would tear ourselves away would be through a bit of friendly treachery.

So here we are, pina colada in hand, while a Trinidad steel band plays... a room 15 stories up over the Atlantic... stuffed lobster tails for dinner last night... boogied at a glitzy nightclub called "Out of This World." Too true, Too much. The cabin seems like it's as far away as the back of the moon. How will I ever thank old Scruggs for a little perspective?

I've lined the front of the firebox with firebrick, added the air ducts and am now building the raised hearth.

Looking back on this week: I'm two rounds up on the front of the fireplace, and most importantly, have discovered how to lay an arch. Friend Mike Andrews loaned me a book on rockmasonry which has enlightened me: I build a wooden form that exactly duplicates the fireplace cavity, then I lay the arch-rocks (all slightly wedge-shaped), building in from each side and resting them on the form until I get to the center and top of the arch, and lay the triangle-shaped keystone to fit in the center and this holds it all in place by wedging. Then you wait until the whole thing dries and sets up before knocking out the form.

So... I'll try it.

Also, Thursday and Friday we began plastering the kitchen walls with a white mortar substance called "Surewall." First I had sheetrocked the room, then nailed chicken wire to the walls for lathing, and then we started paddling the stuff on like stucco. It looks so great I could just holler.

I hauled floorboards in Obie the Construction Bus over to Joe Ryan who has a huge planer in his workshop and together we planed the floorboards out of the Giving House for the new kitchen floor. The whining blades chewed into the dull top of the old pine boards exposing new golden heart pine beneath. In spite of getting blown and covered head to toe with sweet-smelling sawdust, it was thrilling to witness the rebirth of those boards.

Saturday, Sept. 3, 1977 — More Surewalling going on in the kitchen. Also today I built the arch form out of 2x4s and plywood, installed it, and have started in laying the arch. This is fun!

Mike Thompson, my best old buddy, has agreed to hire out at $3.50 an hour, and help us get in by Christmas. His carpentry skills and love of the rustic will make these last few months a joy. This week he came and hung the two living room doors using the old frames and doors from the Giving House. And is he good. He's teaching me a thing or two about "finish carpentry." It's a whole different discipline from framing. Everything has to be spit-spot.

Saturday, Sept. 10, 1977 — Late this afternoon I finished the fireplace. But the really wondrous moment came when I removed the arch form by sawing off the legs — et voila, there stood the arch — with no visible means of support.

No one around to witness what seemed to me the small miracle — so I whooped and danced around the cabin floor. The rock had actually done my bidding. It strikes me as slightly incredible — as in: beyond belief — that I, Mr. Zero, could construct such a loveliness.

I built a small christening fire in the fireplace and then raced outside to witness smoke lazily drifing out of the chimney for the first time.

Sitting on the hillside above the cabin with early autumn's dusk coming on and the crickets winding down, I was filled with profound satisfaction at the ageless sight of smoke curling affectionately above a cabin, of hearth and home — and of the two really connected for the first time. To borrow a Rogers and Hammerstein line, it was "a cliche coming true."

Sunday, Sept. 11, 1977 — Carlton Mason came and built two more stair treads — tricky ones in the arc of the turn so it took a lot of time just to take those two steps. It was a single-handed job, so I installed two stationary windows in the upper gables of the kitchen — and then like a dumdum cracked another window as I was trying to shove it into its tight berth. Too late, I'm reminded of Carpentry Rule No. 48: "Don't Force It, Son."

160 *Hogwild*

Saturday, Sept. 17, 1977 — Maggie and I decided it might be worth it to have the logs and the old wood doors on the north wing sandblasted, so for two days, Charles Martin of Union Mills, outfitted like something out of a sci-fi thriller, stalked around the bathroom and Selena's room blasting the wood down to its honey-golden surface. Well worth the $40 — though we'll be sweeping up sand for weeks.

Sunday, Sept. 18, 1977 — Everything is snowballing in a wonderful synergy: today while Carlton was finishing the stairs, Mike worked on door frames and I continued plastering the kitchen walls with Surewall.

Once the stairs were complete, it was such a pleasure just to walk up and down, looking rather foolish maybe, but it's as if I was feeling the sensation of walking up and down stairs for the first time. Carlton made the rise and tread so satisfying.

And on top of that, he insisted on charging us only the $64 for the materials. The two-inch thick white pine treads are stout and very blond. Quite a contrast to all the old gray wood of the logs, but it is not objectionable. We'll stain them lightly and then polyurethane them for protection.

Sunday, Sept. 25, 1977 — David Kirby helped me install the two big west gable windows — a real chore because we had to scale two 20-foot ladders and carry those hefty moms up

Carlton Mason's stairs are being built of sturdy white pine.

there. It was just one of those jobs I couldn't have done alone. Now that they're in, it's lovely.

Today I also completed the Surewall stucco work in the kitchen. Mike and I have also started framing in for windows along the east gable over the loft so we can get our sizes and go ahead and order the windows. Now that the stairs are in, we can put down the loft floor. Mike has been more than worth the $94 I paid him for six workdays. His emotional support has buoyed me up as well.

The final thing, finished on Sept. 10. Am I amazed or what?!

Wednesday, Sept. 28, 1977 — Report from the Home Front on the Occasion of Baby Jon's First Birthday: it has long rankled me how when certain smalltown newspapers (our own included) publish those cutsey first birthday pictures that the yearlings always look so confounded nice. These kids look heaven-sent. Veritable apple-cheeked, rose-petal, cream-fed cherubs caught by the dime store photographer with the touch of a Rubens. In short, not at all like real 1-year-olds.

Slowly the seed of an insidious smart-aleck idea began germinating. What if some parent were honest enough to picture their kid caught red-handed in some typically awful activity. Ah, he'd be striking a blow for honesty then, wouldn't he?

So when Jon's birthday rolled around, I was on the sleuth for the First Birthday Picture that would tell the capital-t Truth. All week I've been watching him extra carefully waiting for the perfect moment. I had my pick.

He unraveled a whole roll of toilet paper, tearing each downy soft segment apart, and we found him sitting in the bathroom in a Charmin little nest of Carolina Blue stationery.

He investigated the cupboard, removing everything from the lower levels, discovering that Kayro syrup and salt shakers could be opened and were some fun to pour together and then mush around on the floor with the seat his diapers.

Other choices included: turning over the cat water, eating cat food, punching holes in the screen doors, playing in the commode if the top is left up, tearing pages out of magazines, eating same, beating on the TV screen, trying to gargle with gravel, and thumping the bejabbers out of my long-suffering, old black Royal (well Dad does it, he makes it look like so much fun).

But this weekend I finally found the truly accurate image of King Kong Jr.

We were working at the cabin, Maggie sanding walls in the bathroom and I installing windows in the kitchen... when we both noticed independently that it had gotten real quiet in the living room where Jon was supposedly ensconced in his walker. Feeling that prickle of parental premonition, we both

We caught him eight feet off the floor and still climbing.

Down in our swimming hole, mother and waterbaby.

raced for the living room to find to our horror and delight (her horror, my delight) the little stinker confidently scaling the 20-foot extension ladder. There he was, eight feet off the floor and going strong — right over the rock hearth, his face wreathed with a look of bodacious accomplishment.

I had to make Maggie wait until I could grab the camera for the picture. Now that is what I call an honest representation of a 1-year-old manchild.

Saturday, Oct. 1, 1977 — Hogwild is growing! Today Stuart married his long-time girlfriend, Cindy, a sweet Rutherford County girl with long blonde hair, an easy smile, and laughing eyes. Jim Walther helped perform the ceremony on the lawn outside Bill and Jeanette's Big House. The marriage processional consisted of Will Hogwild and the Rocky Broad Boys on guitar (Bill a.k.a. Will Hogwild), mandolin (Andy Morgan), fiddle (Joe Ryan), and bass (me) playing some sweet old country string-band waltz.

A dance and a covered-dish supper followed on the grounds, where most of us ended up as the night wore on... I don't actually remember much about the event at this point.

But I do remember that Jim Walther, he, late of the ministry, was asked to give a blessing before we chowed down. Jim (whose God must have a pretty well defined sense of humor) spread both arms and announced with a straight face that this was an "Old Cherokee Blessing." And as a hush fell over the crowd in the yard, Jim said something like (and I paraphrase freely here): "Yaaaah -- nahpooky, ting-tang, wall-awalla - bing-bang, ooh, yeah-yuuyh..."

And since that seemed to satisfy everyone, we fell to eating, drinking, making music and merry.

Wednesday, Oct. 5, 1977 — I was bending nails at the cabin this afternoon when the whine of a low-flying airplane disturbed the valley's peace. It was circling slowly over Hogwild as if somebody up there was really interested in us.

Supposing it was a carpetbagging land developer casing out the joint, I ran out in the yard and shook my fist at the sky. "It's mine and don't you forget it," I yelled.

Then directly, the farm bell down at Bill and Jeanette's began to toll. Then I heard Jeanette shout, "I need help!" Grabbing my snakebite kit I took off licketysplit up out of Sourwood Holler, only to find Jeanette calmly painting the porch rail. But this didn't settle my nerves — the last time she called for help I found her just as calmly refinishing furniture yet with a big gash in her foot where she'd fallen on a rusty bucket edge — so I knew the lady could be stainless steel.

"It's okay," she said. "It's just a game warden stuck down on the lower road," she shrugged. "I tried to tell him."

The lower "road" is that trail to Burwell and Stuart's land reserved for four-wheel drive vehicles, tractors and the trail bikes Burwell and Stuart sometimes use. It was not designed for a low-slung Plymouth Fury which was now stuck up to its arm-pits.

The game warden was polite and chagrined. He said he'd been alerted by the circling spotter plane that two figures had been seen below looking suspiciously like illegal bow hunters skulking about. After his car got stuck, he'd gotten out, his gun at ready, and was preparing to sneak over there and nab the poachers red-handed. That's when Jeanette caught up with him and asked him what in Sam Hill he was doing.

About that time, Burwell and Stuart, who heard the alarm bell too, came on the run. And when it turned out that these two fellows were the "two suspicious-looking characters," the game warden was fairly speechless. More so when they helped him push his car out of the mud.

And I can still see the look on his face when he found out the two "outlaws" had been doing nothing more than digging their outhouse.

I told the warden it would be a perfect shame for a man to get shot for digging his outdoor privy. He quite agreed and left post haste.

Burwell and Stuart on the porch of the little cabin.

15

In Which the Sweatmore Construction Co. Gets Fancy

"It is better never to begin a good work than, having begun it, to stop."
— *Venerable Bede*

"Blessed is he who has found his work; let him ask no other blessedness."
— *Thomas Carlyle*

Wednesday, Oct. 12, 1977 — It all boils down to a matter of style. Style of living, the trappings you choose to outwardly manifest those values you hold to be self-evident. Even as Mike and I work on what seems to be the skeleton of the cabin's interior, we are confronted with design problems that demand a certain final stylistic understanding on our part. We have to have a shared vision of the cabin's overall look ultimately in mind.

Where you live (the neighborhood, the kind of house it is, the way it's decorated) says so much about its occupant. A house reveals the inhabitant to the keen observer. It is a statement of who you are, and it shouts that proclamation whether you want it to or not.

We have learned this simple lesson by living in the Blue Shed for three years — where to each other and to friends we steadfastly maintain: "We don't really live here, we're just camping." We say, "Just wait — you'll see the real Jock and Maggie when we get the cabin finished. Then, you'll see our true style; then we'll be living where we live for sure."

Abruptly, we have discovered we have been "camping" at the Blue Shed longer than we have ever "lived" any where else in our 10 years of married life. The irony is not lost upon us, and it hastens us onward.

The Blue Shed, quite on its own, has taken on a design look, then,

The Sweatmore Construction Co. 167

that is something of a practice run for the cabin... refinished old pine floors and high ceilings, wainscoting and wallpaper, woodstoves and lots of bookshelves, Tom Cowan watercolors and my black-and-white photographs, oriental rugs and pottery ware, exposed beams and energy-efficient windows, bare logs and the latest in kitchen fixtures and appliances — it's a combination of things both old and new. We think we have a new word for it: "rustslick." A melding of the best of the old — the rustic, with the most modern of things — the slick.

So today Mike and I leapt into an ambitious project that required quite a bit of design skill. First, we floored the loft, which was simple enough. Then we built a frame for the divider between the loft and the cathedral ceiling. This involved a lot of what Mike calls "nose-picking time" — that's what Mike irreverently calls my creative carpenterial design meditation. We came onto this idea of facing the loft divider with the old exterior tongue and groove from the Giving House. This stuff is 12 inches wide, at least an inch thick and weathered to a silver-gray.

Our design scheme is based on the angle work I've started in the bathroom. Mike likes it so much he suggested we use that as a thematic device, and do a "V" shaped angle pattern.

Mike and I frame the loft, adding silvered exterior tongue and groove from the Giving House.

Working at 18-foot heights with exacting 42 degree angles and old wood is no simple feat, so I was really glad I had help on this project — and when done, the wall turns out to be a work of art. It is just miraculous. When you first walk in the room, its sunburst lines force your line of vision up, and with it, your spirits rise.

You could hang nothing on these walls and there would be art there yet.

Saturday, Oct. 15, 1977 — We have wondered about a new road running up the creek valley and so have employed a bulldozer operator to come out and look at the situation. As we have explored up the Silver Creek valley above the cabin we discovered an old roadbed that paralleled the creek, and then petered out as the incline started up. It seems to be a natural back entrance,

The other reason for this road is that Bill seems adamant about not fixing up Hogwild Blvd. I believe he actually likes it inaccessible, in fact, prefers it that way. But in bad weather, that means: impassable. My problem is that working on the newspaper, I have to be able to get in and out, so maybe the best way to handle it diplomatically is to be responsible for my own access. Hence a new road on my land that bypasses the worst of the mudholes on Hogwild Blvd.

We had the new road dozed today, and the only problem came when the bulldozer operator insisted that we didn't need a pair of switchbacks I said were necessary. He said the switchbacks would make the road too steep, but I maintain just the opposite to be true: that without them, the road would have to charge right up the mountainside to get out of the valley. He said not to worry, and plowed on. How do you argue with someone sitting above you with four tons of growling steel in his hands?

He had his way, and I reckon it's okay. What do I know about roadbuilding? The results are awfully steep in the worst place. But on the bright side, the road as it runs along by the creek is plumb wonderful; the new road opens up the creek valley to the cabin and makes us feel our land is accessible in a way never before realized. We are "tickled witless," as Gene Ham would say.

Also, the dozer man shoved around the front yard, upon Maggie's direction, leveling and extending our yard; and I have to admit, this is something I should have done years ago.

Sunday, Oct. 16, 1977 — Having finished the loft wall with the angled barnboard, Mike and I now have built the closets in the loft; one side for Maggie and one for me. Today we did the finish work on Maggie's side, complete with neatly angled doors and curly-cues that took a bit of chutzpah to do.

Now I'm ready to do some finishing touches on the chinking around the loft logs, and start thinking about the main room's ceiling. Folks tell me that two-inch-thick Styrofoam (R10 to R15) needs no ventilation since it doesn't have a vapor barrier, and that's the way to insulate the ceiling. Then, just use fiberboard, lightweight and cheap, made by Celotex for under three bucks a 4'x8'. You can't beat that.

Wednesday, Oct. 19, 1977 -- The sun loafs down behind cabin eaves. A breeze stirs dry-handed oak applause. A droning airplane makes a sleepy pass to the north. The crickets are slowing down.

I am beguiled by the season of ferment. Something here is about to pass.

The old red dog moans in her sleep in the front porch sun. The light falls golden on the distant bottomlands stretching in a saffron palette to the clear-voiced river.

The cumbersome mud daubers are in a mild state of panic, inspecting each chinking crack, each nook and niche in the shade-splashed rock chimney. They maneuver awkwardly, flying leadenly like B-52s attempting aerial acrobatics. These dark heavy bombers have no malice for me; their stingers are forgotten. They're looking for winter haven; they too feel the warning chill in the air.

I too am a mud-dauber in the grips of the autumnal juggernaut. Will I have my crack in the chink ready when cold weather comes? If I don't get in by Christmas won't I die?

It's too much to contemplate. I buzz on.

Mike takes a work break to visit with Maggie and Jon, the latter of whom has just turned over a soft drink.

Saturday, Oct. 22, 1977 — Mike Thompson is keeping me sane. He keeps me laughing and thinking, questing and questioning all the time he's out here working for his measly three fifty an hour. That's all he'll take. A better friend no man ever had. Plus, he's a born teacher. What better way to be taught than by a friend?

As we build we banter of nails and Nietzsche, of cabins and Kant, of headlines and Hegel, of boards and bards, of Man's Place in the Universe and of our place in the country.

Thompsonian Doctrine: It doesn't really matter what you do in life, just so long as you feel it is worthwhile. Then, dig in, and do it well.

Though Mike is a couple of years younger than I, he is wise and kind in a way I've never before known in a man. The high school kids he teaches and reaches feel this special aura too. He has make an indelible impression on the lives of hundreds of students at East Rutherford High School for six years. We like to joke that he's the county's hippie teacher and I'm his journalistic counterpart.

He salvaged and restored an old house over in Duncan's Creek on the other side of the county back in '73, showing me indirectly that it was possible for a city boy to come to the country and come to terms with tools and objects and living on the land.

His marriage hasn't worked out and he has recently separated and is living in a shacky little house way back up a red mud road in Sunshine. He drives a beat-up red Chevette that has a peculiar penchant for getting stuck in his and every other driveway.

Mike is a Renaissance man and we love him for his wisdom and his vulnerability, his depth and his talent. He has taught me everything from the value of a miter box to Travis-picking on the guitar; he can run backwards faster than anybody I've ever met, a talent he has displayed in hilarious backwards-running contests. He is equally at home teaching international relations and world geopolitics, philosophy, psychology or African studies.

He is a German scholar and laughs about the good earthy sound of the language. To Mike, the word nagels is somehow infinitely more expressive and onomatopoeic than "nails." Likewise with toogles for "tools." Mike is forever blessing me with his lexicon of new words, neologisms that make me look at the language afresh.

I can't say exactly how it started today — but as we were working, the subject of political geography came up, and because we knew we were both knowledgeable on the subject, a little game began.

"Hand me the 16-penny nagels," Mike said, "and give me the

location of the Bosporus and Dardanelles, and what is their political and historical significance?"

"Here," I said, handing him the nagels. "Critical narrow straits between the Black Sea and the Aegean, politically sensitive because it's Russia's southern exit to the Med. Also the conflict line between old enemies Turkey and Greece."

Pausing briefly to measure a board, "We need to cut it here," I said, "and what's the capital of Upper Volta, what language do they speak and what was its colonial name?"

"Okay," said Mike, squinting at the pencil line on the board and then manhandling the circular saw through the required cut, "Ouagadougou, (pronounced OO-ga, DOO-ga) French, French West Africa."

Mike got a new board. "Name the two ethnic groups on the island nation whose capital city is Colombo," he challenged flatly.

"Tamil and Sinhalese," I shot back, "Country's name is Ceylon." Then it was my turn. "Name the second tallest mountain in the world, and give me both of its names." (Ha, I thought I had him here.)

But a grin sprawled across Mike's face. "K-2. Mount Goodwin Austin," he said and then laughed outloud.

"How in th' hell did you know that?!" I shouted back in mock anguish. Mike shrugged happily and turned right around with: "Name the four main islands of the Japanese homeland."

I had to strain for this one: "Hokkaido, Honshu ... Shikoku and Kyushu... whew."

And so it went... a game of geographic wits played in a log cabin in the middle of 300 Hogwild acres.

Sunday, Oct. 30, 1977 — We have framed the east gable's permanent windows and installed the three sections, awaiting now the big center casement.

Our buddy Andy Morgan got married today over in Sunshine in what has to go down in the history of Rutherford County as the most bizarre wedding ceremony ever witnessed. The groom (short, leprechaun-like, curly red hair and red beard, green laughing eyes) wore white overalls; the bride (Andrea, a sophisticate from Tryon) was attired in white cotton bloomers and an antique bodice contraption that I could have sworn was another undergarment. She looked like something out of *Gone With the Wind*, yet I couldn't tell if she was just in the act of dressing or if this was some high fashion statement. Having been in the woods so long, I assumed the latter. As the afternoon went on, she seemed to acquire more old white cotton petticoats.

But that's not all.

The preacher arrived on a motorcycle with black robes flapping in the wind. The bride and groom were both made up like clowns or maybe it was mime: white face with big red cheek circles. The ceremony took place under a giant cargo parachute draped over a large cedar tree from which they had rigged speakers. The nuptial couple processed to Judy Collins singing "Send in the Clowns." There must have been a hundred of us crazies standing in a big circle holding hands under this parachute while Andy and his bride giggled their way through the ceremony like it was comedy. It made perfectly good sense to go out like that, I reckon; she's an actress and Andy is a make-up artist, clown and mime actor. Weddings are pretty much staged productions anyway, so they figured, why not do it like a production number?

Sunday, Nov. 9, 1977 — It's been two weeks now since we had a couple hundred dollars' worth of gravel and small stone, called "crusher run," put down on the new creek road, and I have to report that it was all been washed away by two humongous floods on Saturday and Sunday.

I have a feeling this was no ordinary rainstorm, but a high water mark in every sense of the word: the Rocky Broad left its banks and flooded all of our bottomlands. Even little Silver Creek turned into a torrent and shoved the footbridge 25 yards downstream.

We are sick about the new road. Obviously we should have insisted on those switchbacks and proper ditching. On the brighter side, the new grass Maggie sowed on the new banks back then has boldly sprouted in spite of the rains.

Meanwhile Mike and I undertook and solved a thorny problem, that of leveling the kitchen subfloor which was intersected by a major support beam, "Big Mama," which actually stuck up an inch above the floor's surface. So we adzed and axed and planed her down, then subfloored the whole place with particle board, and filled the cracks where the new wood met the old floor beam with Surewall, so that when the final flooring goes down it will be relatively level.

We have also installed the last big Andersen casement window in the east loft gable. While Mike has done the difficult finish work on my closet, I have been sheetrocking the east gable, and then Surewalling the interior. Mike has been proving his worth as a skilled carpenter: every single one of the angles in the loft closets are obtuse or acute. And when we ran out of old pine from the Giving House, we used old oak bought from a nearby salvage place for a couple of dollars. The grayed oak doesn't fight with the heart pine as I feared.

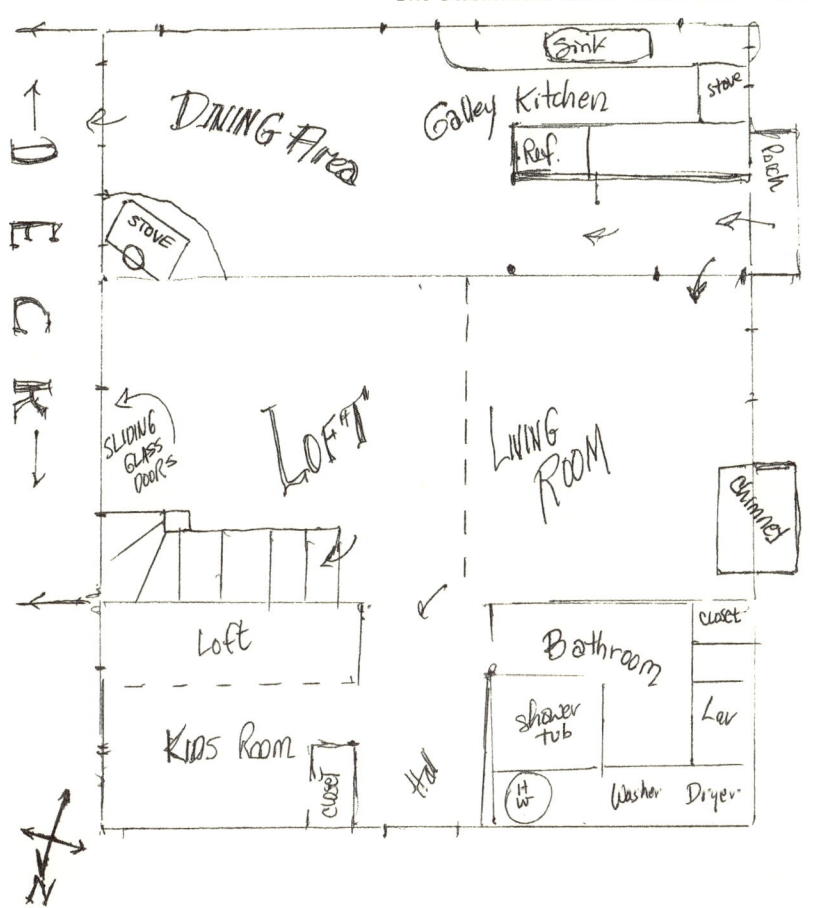

The basic floor plan. Entrance is upper right.

Sunday, Nov. 7, 1977 — Mike didn't come today so I sheetrocked the west gable — working alone on the high ladder with 4x8 sheets of the unwieldy stuff.

After today I have a new definition of eternal damnation: single-handedly sheetrocking the ceiling of hell.

Saturday, Nov. 12, 1977 — No, wait. I may have found a better damnation: single-handedly insulating hell's basement.

But thankfully I had help on this odious task. Mike and I did the kitchen floor insulation in nothing flat — working together and with plenty of crawl space. But the living room was altogether different. The crawl space is about two and a half feet and if we hadn't had each other to cuss at, joke to and keep sane, I do believe I might have gone plain bonkers in that dark, dank narrow cave with the infernal itchy fiberglass getting all over us and the sweat just pouring off.

But Mike is such a trouper. Together we go into sort of a carpenter's overdrive, and our hyped energy carries us through the task like it's some kind of joyful challenge. Maybe that's not what the Beatles meant when they wrote "I Get By (High) With a Little Help From My Friends," but that's what the song means to me. The cabin stands and grows as a lasting testimony to so many true friends whose imprint and signature I can read literally everywhere I turn to look.

Wednesday, Nov. 16, 1977 — In order to try and finish this winter (Christmas is looking more and more like a long shot), we have elected to get a building materials loan of $3,000. The final push is going to require that we have that financial resource readily in hand. Up to this point we have tried to pay as we go, but now that we're in the home stretch, so to speak, we have to be able to buy things quickly.

For instance, today we drove over to Rick Pratt's solar store in Tryon and stuffed the van full to the gills with $300 worth of two-inch thick Styrofoam sheet insulation for the ceiling.

Wednesday, Dec. 7, 1977 — I am making my final entry in the well-thumbed, dark red-backed journal Maggie gave me three and a half years ago. Funny, but I'd always assumed I'd finish the cabin on the last page of the "Yes Book." But instead, it feels like only a first installment. A new, burlap brown journal awaits more adventures. A fresh page beckons. The continuing saga of Sourwood Holler is something I may always be writing.

Thursday, Dec. 8, 1977 — Plumber Leroy Sims finished the rough plumbing Sunday. We paid him $439 for the lovely copper work he did throughout the house. I have been hard at it installing the Styrofoam ceiling insulation — a simple and pleasurable operation. I measure the stuff, mark it with a pencil, gently saw it with hand saw, and then just cram it between the rafters where it is held in place by wedging.

Also on Sunday Mike and I started the final flooring in the kitchen with the old Giving House tongue and groove. When we got halfway through we saw had just enough to finish the kitchen but not enough to do the front hall as well. That area (4'x8') will require a different set of salvaged wood.

The problem arises from inconsistencies in the salvaged flooring from the Giving House where the floors were cut, milled and laid in different times, and therefore they were of different widths and thicknesses. That requires me to use them faithfully in this reconstruction.

Thursday, Dec. 15, 1977 — We finished the kitchen floor on Sunday, and today I worked alone shingling the exterior of the east and west gables around the high windows. On the west gable it went very quickly as I did a straightforward pattern. But on the east side overlooking the deck, I elected to do a sunburst pattern that radiates down and outward from the apex of the gable.

This required trips up and down the ladder for every single shingle to be measured and custom cut. But the result is eminently satisfying.

Sunday, Dec. 18, 1977 — Before Mike left yesterday on a much-needed two-week vacation to his parents' place in Florida, we framed and sheet-rocked the front hall divider, and hung another old door from the Giving House so that the cabin will have a foyer with a cold air baffle.

In a fit of orderliness, Maggie and I decided a major clean-up day was in order, and so we swept sawdust, re-stacked salvaged wood and organized tools. By the end of the afternoon, the place looked like it made a little more sense.

Sunday, Christmas Day, 1977 — A year ago today I said we'd be in the cabin by this Christmas, so I have to be disappointed that it didn't happen... disappointed in myself and ashamed that I've let Maggie down. But shoot — we couldn't have worked much harder.

So I'm trying to reconcile the failure. Or at least what feels like failure. Again, I have set myself up to be tyrannized by time. So... what lessons do I draw on this snowy Christmas day? It must be that I should not worship time frames so strictly. Things must be allowed to happen in the fullness of time. I should instead rejoice in the good; the fact that I am not sacrificing quality workmanship for time or speedy completion.

Thursday, Dec. 29, 1977 — Working in the cold, with a cold, I today all but finished the high east exterior shingling on the gable. It's a good singular project that I can piddle on while Mike is gone. Today the temperatures were in the teens and really too cold to work, and I was too sick to work, but I did anyway.

176 *Hogwild*

Saturday, Dec. 31, 1977 — Another cold day, but what fun. My young friend Jimmy Grogan, editor of the school newspaper, came out and wanted to help. So I figured out a project for us to do — which was building the flat kitchen hearth for the woodstove that's going in the corner. We need an insulated rock platform on the floor and insulated backing up against the log wall as well. It took all day to lay the five-foot-wide semi-circular hearth. Jimmy mixed mortar and I showed him how to lay the big, broad flat rocks from Silver Creek. It looks wonderful and I enjoyed teaching this gifted young man such an old craft. He's a Morehead Scholarship nominee to Chapel Hill and I bet he gets the prestigious scholarship. And if he doesn't, he can always be a rockmason.

Tomorrow is the first day of 1978, which marks the fourth year of work on this cabin.

I find that difficult to comprehend.

The kitchen hearth base is complete; Jon takes measurements.

16

In the Home Stretch

*"You know the nearer your destination
the more you're slip-slidin' away."*
— *Paul Simon*

Thursday, Jan. 5, 1978 — I am going to finish this dadblasted albatross of a house this year or else it's going to be (with apologies to Kurt Vonnegut) *Welcome to the Monkey House*.

On this first working day of the new year, I gathered flat stones and laid the first vertical course along the walls for the kitchen hearth. We are anticipating a big double-doored Fisher woodstove from Russ Scruggs.

Alone with the cabin all day. No one came down. The rock rising surely from the floor. A wind blowing through the open doors on a chilly January day. This piece of rockwork is the easiest yet most free-form I've done. The base is a triangle with a rounded front, and the sides are fan-shaped. The design happened on its own accord. One more day

Down at the Big House: we celebrate January birthdays.

and I'll have it done.

Celebrating a pair of Capricorn birthdays, last night we feasted at Bill and Jeanette's in fine style. Bill turned 30 this week, and Jeanette, I think, is 27 next week. Maggie and Jeanette cooked one of their typically great meals, served with wine and the roaring of Bill's huge Fisher "Grizzly Bear" fireplace stove, followed by music of our own making by Will Hogwild and the Recycles (Bill, Joe Ryan and me).

Saturday, Jan. 7, 1978 — I can't believe it. I hear third-hand that Mike Thompson is not coming back from his Christmas vacation to Florida. This is very strange indeed. I am truly boggled by this development. What could have happened to him?

Apart from the personal considerations, this is a set-back for the cabin, which was a couple of months from completion so long as there were the two of us banging away out there. But now... I just don't know. It feels like another six months at the very least if I have to do it single-handedly.

Reckon I've become spoiled. Having Mike has shown me that some jobs are just ludicrous to try alone. But then, I did it for three and a half years, and by damn, I'll do it again. Besides, I've been doing well enough working alone these last three weeks.

But today I've been in shock over Mike. Rocklaying on the kitchen hearth has gone very slowly and will require another day to finish. What I've done so far is pretty fine.

From there I don't know what to do next. I've got a couple of months of little doodlysquat things that need to be tied up:
— Build a front door from scratch.
— Front hall final touches: Surewall and wainscoting from barnwood.
— Get triple-wall stovepipe for kitchen woodstove — install.
— Plane flooring for living room floor — install.
— Custom kitchen, bathroom cabinets —have them installed.
— Ceiling (Celotex or U.S. Gypsum insulation board).

Friday, Jan. 13, 1978 — Deep in winter's grasp tonight. It has been so bitter cold that I can't work with the mortar, so the kitchen hearth still awaits a warm day. Maggie is away singing tonight in snowy Asheville.

Taking advantage of the cold to run errands, Maggie and I went to Tryon yesterday and got the triple-wall steel chimney from Rick Pratt's wood and solar shop.

We also visited potters Claude and Elaine Graves who live outside Columbus in a dome house made of concrete. I needed this inspiration

Homesteading neighbors Claude and Elaine Graves of Columbus and their "big gray elephant."

because he's been working on it I believe he said for five years. Now that I'm in my fourth year, I need all the encouragement I can get.

The big difference between us and the Graves is that they've been living in their house while they've been working. While this has been stressful to them, (especially with baby Zeke tottering around the construction mess) it has also been satisfying. We committed early on to not trying to live in the clutter of housebuilding and I think I'm glad. Yet talking with the Graves made me yearn to be in, no matter what state of completion the house was in.

Maggie has engaged the Graves to make custom-designed tiles for the kitchen: green vines and Delft blue birds on white background. It's very exciting to get to the point where we're finalizing on the finishing touches.

Sunday, Jan. 15, 1978 — At last! Two weeks after starting, I finally finished the kitchen hearth. This was the first day in ages warm enough to work (even while Wednesday's snow still rested on the rocky shoul-

The kitchen hearth is complete. It's the smallest yet most artistic rockwork I've yet done.

ders of the new chimney, which in itself was a new sight).

Finishing the hearth was deeply satisfying for me; there's just something so innately physical about rocking. It is an ultimate tactile experience, combining the best of art with weightlifting. The thing itself is an authentic reflection of the land: rock, sand, and water — all came from Silver Creek.

I spent a lot of time selecting, sorting, and trying every imaginable combination of rocks until I was content with my final choices to finish the top row. By sundown it was finished — and just as I completed it, the last golden rays of a winter sunset blasted through the kitchen window like a follow spot to highlight the new rock star. The paparazzi rock fan takes pictures.

Yesterday we sank $160 in light fixtures, a bathroom sink and commode. Because lighting is intrinsic in establishing mood, we have been careful about these choices: plain white spheres for the bathroom and back hall, a simple four-socket indirect light for the sink, an old-fashioned coach light-style front porch light, spotlights for the kitchen galley to be mounted overhead on the Big Daddy central beam that runs the 22-foot length — and for the front hall we are still looking for a perfect hand-woven basket to use for the lampshade.

The bathroom fixtures back up our "rustslick" design motif: a white bowl sink, fixtures that are simple but Danish modern in style, and a low-slung Star-Trekky looking commode.

Saturday, Feb. 4, 1978 — Knocking out doo-dah projects left and right now: finished the final shingling of exterior high east gable, insulated and sheathed like a barn door the old hay loft opening of the east gable, decided where to put kitchen stove in such a way as to make the kitchen galley wider.

And finally, I filled in the last section of subflooring in the living room which I'd left open by the big raised hearth while Mike and I were insulating the basement. This last job was dictated by safety considerations after a narrow miss last week: Selena had her friend, Teresa Brewer, come over to play. Scampering around the cabin, Teresa ran through the living room — and suddenly disappeared through the hole in the floor — just as magically and terribly as the entrances and exits of the Wicked Witch of the West from Oz.

The fall may have been only a three-foot drop, but it was between beams, concrete blocks, and a rock hearth. I was terrified as I saw the hole swallow up the child. Then, there was her little face, poking back up out of the hole, with her eyes big and round in amazement. Aside from a scratch or two, she was unscathed. We took that as fair warning to close up the trap door.

Sunday, Feb. 5, 1978 — A great day! In spite of odds, we got the Fisher Grandpa Bear down to the cabin and onto its hearth in the kitchen — a job I really dreaded because the big monster was one chunk of cast iron and steel.

We took the newspaper van over to Lattimore in the next county where Russ Scruggs lives. He had the stove waiting for us on his porch, and using a mover's dolly borrowed from Joe Carson's furniture place in Rutherfordton, we got that thing into the van with minimal hassle.

Then back at Hogwild, we found Bill and our prospective cabinet-maker, Raymond Lee of Country Cabin Collection. So we had plenty of muscle. After backing the van to the porch, we wheeled the stove into the house and up onto the hearth, where we placed it catty-cornered. And the fit is just right, and not an inch to spare, either. Jubilation.

Then Ray and I worked up the measurements and estimates for the bathroom cabinets Maggie has designed. His estimates were quite modest I think for the kind of custom quality work his shop does. He promises to have the work done in a month, at which time we'll deliver kitchen drawings to him and they'll go to work on that. Et voila. We are beginning to cook.

Saturday, Feb. 11, 1978 — Are we being treated to an early spring this year? It sure feels like it. We've been blessed with three wonderfully warm days here, and in spite of my fighting the annual case of the nasal gunkywunkies, I've gotten more done: finished west high gable shingling, finished sheetrocking in entrance hall, and best of all — insulated and sheathed east kitchen exterior gable at last. I say "at last" because it was done backwards. I'd sheetrocked and stuccoed the interior before completing the exterior. I have to shake my head and wonder at my own bald-faced self sometimes.

The dry warm weather has also allowed us to drive down the ruination of the creekroad (dubbed "Four Wheel Drive" because that's the only type of vehicle that can scale it). We'll have to get that "road" reworked this spring.

Friday, Feb. 17, 1978 — Recovering from a bout with the flu. I am all but finished with papering and shingling the east kitchen gable exterior. The work is rewarding and attractive, tying the wings of the cabin together stylistically with the central log core.

Saturday, March 4, 1978 — Excitement! The bathroom cabinets arrived and we installed them today. They are simply fantastic. The boys at Country Cabin Collection built them to our specs: natural pine, cut nails, black hardware, and overall stained Minwax's Puritan Pine and then finished with a low gloss. They are worth every bit of the $408.

Sunday, March 5, 1978 — Good old Joe Ryan worked with me well into

Windows are in, high and low, and I've shingled the gables. Maggie is painting the trim a Wedgwood blue. Now with the leaves gone, we can see the chimney's full beauty.

the night getting the kitchen woodstove vented. We used handsome black eight-inch stovepipe that winds around artistically to the ceiling connection of the steel chimney which itself took all afternoon to install. We lighted a ceremonial fire to test the new stove's draw, and the christening cheered the kitchen.

Sunday, March 12, 1978 — Spring trips shyly into the Holler and with it comes a renewal of the blooming log cabin builders. Today I attended to hanging and finishing the hall door to the kitchen, an old thing from the Giving House, while Maggie painted the windows a color we call Sourwood Blue which is Wedgwood Blue to anyone else. Late this evening while a warm dusk deepened the spring shadows, I finished framing the prospective front door which I'll be building from scratch.

Saturday, March 18, 1978 — Maggie is still floating from the exhilaration of her one-woman jazz show, *Maggie Lauterer Sings*, that she produced and starred in at the community college. It was a marvelously successful jazz/pop concert. Two standing ovations.

Today, back in our log cabin world, Bill and I are equally high, for

184   *Hogwild*

Bill and I have just put down the new (old) living room floor.

we have laid one fourth of the living room floor. With his expert help, it goes quickly and easily. Meanwhile in the kitchen, Jeanette is sanding and refinishing that floor. What energy!

B&J have really pitched in here and have gotten us rolling again. Of Mike's departure and rumored new career in Florida I can only wonder with incredulity at the switch from log cabin builder to stock broker. I'm sure the next time we meet he'll help me understand how the transition was a logical one.

Saturday, March 25, 1978 — After three good days flooring, Bill and I finished laying the living room floor, while outside a spring shower accompanied our hammering. In the kitchen, Jeanette continues to caulk and sand down the old floor Mike and I laid in December. To ward off a spring chill, we had fires going in both stoves today, and it made the place seem downright homey.

Thursday, March 30, 1978 — Manna from Heaven? Bill and I have enough flooring left over to do the front and back halls as well. Amazing. And I was worried that there wasn't enough flooring to do the living room — and now we have plenty for the whole house. Something about the parable of the fish and the loaves comes to mind here. Thanks be.

In the Home Stretch 185

The power company folks came today and trenched for the underground electric cable so we won't have any unsightly lines coming into the cabin. And when they were done, these good old boys walked around the house looking at the place with open admiration. One fellow said the line I hear so often that it rankles: "Gonna be nice when you get it done."

Friday, March 31, 1978 — Jeanette is doing the most lovely work: four coats of polyurethane in the kitchen has turned the old floor to gold. And then she stained (Puritan Pine) the new wood of the stairs and the loft, and then finished those areas as well.

Meanwhile, I have framed the interior of the little window by the living room hearth, using the last of the old sandblasted heart pine. This took much chiseling of the surrounding logs to get the fit just right. It is my best finish carpentry work to date and I'm deeply satisfied with the results.

Wednesday, April 5, 1977 — The 47 Things. They stare up at me from the legal pad. They're what I have left to do. Everything from building the front door to doing the ceiling to running the water line. Each one is a separate personality yapping for attention and howling my name. I am like an ol' mama hound with too many pups and not enough fixtures.

Friday and Saturday, April 7-8, 1978 — I have learned the oriental art of compartmentalization. Isolate the problem of the highest priority, shut out the clamor of the others, and complete the selected task in utter concentration.

Today was a truly (Hog) wild day with a major work force doing separate jobs all over the cabin. I was at wit's end to coordinate it all: Bill and I finishing window trim in the kitchen while Maggie and Jeanette worked on floor finishing in kitchen, loft and stairs. I was accused of being grumpy. Well hello Jack, with the end in sight and after four years, I reckon I am feeling a little zoomy.

Sunday, April 9, 1978 — Sometimes I get the feeling I'm more of a participant in this pageant than I am its producer. The interior of the cabin is assuming its own marvelous identity: in the front hall the chevron design wainscoting of old barnwood echoes the pattern of the living room loft divider, and mirrors again the angle-wood design in the north wing as well.

We put in a good ten hours here today, completing the front hall wainscoting, doing final trim on the windows and bathroom door with salvaged oak and pine.

Saturday, April 14, 1978 — Cosmic days. Three in a row. For a reason known only to Abba, Allah, Baba, Buddha and maybe even Bubba, I felt like tackling the front door.

To begin the interior side, I nailed Giving House tongue and groove onto a plain core of three-quarter inch plywood in a simple vertical pattern.

But for the front I wanted to do something more artistically articulate. And the thought occurred that the front door ought to be like the cover of a book, that it ought to alert the reader or the entrant that what was about to follow inside would be a reflection of that frontispiece.

So to establish the mood, I did the front door out of old pine in a diagonal cross pattern. Mind you, I've never done anything like this before, so it was a pretty outrageous thing to attempt. When it was done, and it far outstripped my wildest expectations, I was filled with a profound sense of accomplishment — both as a finish carpenter and an a graphic artist.

Saturday, April 15, 1978 — My generation and those younger take three vital life-supporting factors for granted: heat, light, and water.

Folks raised in the Depression times and earlier still marvel silently each time they walk into an evenly-heated warm house, flick on a wall light switch, or turn on an inside faucet. After four years of heating exclusively with wood, I've learned not to take heat for granted. But I still admit to expecting electric lights to respond without my thinking of the miracle. And this weekend I got my initiation into the why's and wherefore's of water.

Before Hogwild, if someone had asked me where water came from, I probably would have said: "out of the faucet, of course."

I began dreaming about a gravity-fed water system four years ago when we discovered from the original survey maps that Bill and Jeanette got their water from an old double-headed spring high up on our land — fully 2,000 feet from their house, but only 750 feet from our cabin. The fact that my spring had been feeding their house with water since 1915 attested to the spring's prodigious capacity. Indeed, the spring's overflow created what we dubbed Spring Creek which forms the narrow verdant gorge between Slickey Mountain and Laurel Mountain. A tiny path winds up the wildflower-strewn hills and under the groping mountain laurel thickets to a large rock-lipped spring which was covered with ages of leaves, sediment and a giant, round, once-red Coca-Cola sign.

Yesterday I turned into a sherpa and made repeated assaults up the narrow creek gorge, lugging 50 pounds of Sacrete, joints, clamps, assorted tools — and most ungainly of all — 600 feet of plastic one-inch piping. At the top, there beside Bill and Jeanette's spring lay my reward: another pool — deep and clear and still. Wallowing around in the silted bottom were an assortment of spring critters — tiny wriggling salamanders, snails, and one granddaddy mudpuppy.

With local rock and my concrete mix, I built a retaining wall to form a dam and make the holding pool deeper. And through the dam I imbedded one end of the pipe, encasing it in window screen, and covered that with a perforated tobacco can to act as a primitive filter and salamander-baffle.

Then I stumbled backwards down the middle of the creek with the giant coils, unrolling the piping, and connected the 100-foot lengths with joints and clamping until I got past Selena's Playhouse and to the low point in the waterline where it crosses Silver Creek before rising the last 100 feet or so up to the cabin. There I stopped for the day.

Sunday, April 16, 1978 — "Water water, everywhere,/ Nor any drop to drink..." *The Rime of the Ancient Mariner* comes to mind today, for at first it seemed like even with the plenitude of water, I would have none.

This morning after the concrete dam had hardened, I went up to the head of the waterline at the spring and connected the pipe to the outflow pipe — and nothing happened. I don't know what I expected; I guess I just thought the gravity would send the water a'gushing. But I still had more to learn about hydro-engineering.

Then it occurred to me that maybe it needed a gentle nudge. I clambered down the rocky creek to the first 100-foot junction of the pipe, disconnected it, and took it to my lips like I was about to give a blast on the world's largest Alpenhorn — but instead began sucking 100 feet of air out of the pipe in hopes of getting the suction going.

When I was about blue in the face I heard it coming like a highball express — rushing and jostling and chuckling to itself down the pipe. Then *whoosh*, the springwater exploded through the pipe, all over me, and sending the pipe writhing like a live thing. Joyously, I jammed the two sections of pipe together, getting sprayed thoroughly as the pressure fought the union, clamped the pipe down, and ran with sopping tennis shoes like a wildman down through the middle of Spring Creek, oblivious to the leg-clawing branches and face-slapping vines toward the base of Spring Creek where, long before I got there, I could hear the geyser — a gusher that had blown off the gate valve I had loosely clamped to the end.

The awesome power of gravity and water pressure astounded me. From that meek bubbling pool up the cove had come this thrashing powerhouse. I whooped and cavorted about like old Dan Gunn of *Treasure Island*; a bedraggled figure in wet cut-offs and raggamuffin tennis shoes dancing about in the gushing spray.

Later, after calming down somewhat, I attached the gate valve at the low point at Silver Creek, turned the flow off, and ran the last 100 feet of piping uphill to the cabin and put a proper end valve on, wondering to myself: can water really run uphill?

Then I turned the whole system on, and Kowabunga, Buffalo Bob. The whole shebang worked like gangbusters — water all over the place!

I make this vow — that when it's all said and done, piped into the cabin, that when I turn on the spigot and spring water comes a'gurglin' — never again will I take for granted that seemingly simple miracle.

Sunday, April 23, 1978 — After Maggie and I sanded the living room and hall floors for what seemed an interminable time yesterday, and after having blown through fifty bucks worth of sandpaper, we finally finished. Did we save money? I don't know — but I do know we by Jiminy did it ourselves. The reward came when we put down the first coat of finish, and that clear polyurethane acted like a stain, and to our surprise the sanded, ash-blond old pine turned to a deep red cedar color — much darker than we thought it'd be, but lovely in a woody, resinous depth that is already growing on me.

Friday, April 21, 1978 — Still burnt out from an all-nighter Tuesday putting the paper to bed... nevertheless today I built the closet in Selena's room, and best of all, a tiny loft in her room that will eventually be Jon's bed when he's old enough to climb up and down safely. It's neat and strong, built out of salvaged culls from several flooring projects.

Saturday, April 29, 1978 — A day of serendipity. After bonking my head for the fifty-eleventh time on Selena's low door, I came to a revelation: (first of all, when you're bald, that's a capital-b Bonk) that when a feature of a house keeps hurting you then it's not built right.

And so without further ado, I ripped the door apart. What was that line from Ecclesiastes about "a time to build up and a time to tear down..."?

This was one of those times. I'd built the door using a short door from the Giving House and a nice old piece of plate glass that I can remember my mother mounting on white fire bricks and using as an art deco coffee table. I can remember this glass table from earliest childhood, hence its sentimental value.

Later today, while building the exterior north "door" (in quotes because it's a permanent wall and just made like a false door) at the end of that the north wing hall where Jon's crib will go, I had a vision.

A window. It would be such a shame to close off light to his little room... but where to get an inexpensive window just the right size without going back to town and ordering one...?

Then "the Eureka factor" kicked in with Maggie suggesting that I use Mom's little glass coffee table for Jon's window. So, that which I had just destroyed gave me what I needed to create something new.

After framing it in, I plywooded and tarpapered the little false door, and finished the exterior with a fanciful pattern of salvaged pine angled this way and that. The new construction again mirrors the unusual angled effect of the front door. I've got to admit, I got pretty high over building that little door.

Using that "builder's high" to keep my momentum going, I ran down to Bill's where I had some plywood stored, and carried two 4'x8's back on my back down to the cabin. I spent a wonderful evening sheathing the back hall in its entirety while a full spring moon floated up over Slickey Mountain like a grinning hot air balloon, and the April night was thick with the songs of lovesick whippoorwills.

I get fancy on the hall exterior false door.

17

Moving In At Last

*"Perseverance is a sign of will power.
He who stays where he is endures."*
— *Lao Tsu*

Thursday, May 4, 1978 — It is approaching graduation time. The time frame is too obvious for me to miss the comparison. I have spent four years at Hogwild U. and this is definitely my senior year. Just one more semester and I believe I would have hit the wall. I must finish this dissertation of wood, rock, and mortar by Commencement. The concept is very Zen: that I must finish so that I can "start."

The metaphor works. That first summer session of 1974 I was so naive and bright-eyed. I thought I could skip a couple of grades and get done in three months. As a freshman I thought a toe nail was something you clipped, that a plumb bob was a character on *The Waltons*, that it was impossible for water to flow uphill, and was so ignorant of one tool that I proudly "invented" mine of sheer need — only to find two years later that any dingdong carpenter or kid who took shop in high school knows what a T-bevel is.

Then, depressed by the immensity of the project, I glided cynically into a second year, becoming a proverbial wise fool along the way. I found out that the more I did, the less I knew.

Along the way I met classmates at this land-grant college who in turn helped me cram for pop quizzes. Tutors were everywhere. Distinguished Professors of Carpentry Byers, Ratcliffe, Thompson and Ryan guided my studies. Visiting Scholars in Rock and Woodwork Kaltreider, Mason, Morgan and Kirby helped me hone my new-found skills.

As so often happens at college, the junior got his act together. Realizing what my major was, I dug in with renewed focus and intensity. I was charging so hard that I scarcely noticed when I slipped into the comfortable maturity of the senior year.

This spring I have hit my stride and it feels good to red-line it. Cramming has become a way of life. Finals loom before me. I've got major exams in every room of the cabin.

I wonder what my diploma will look like?

Sunday, May 7, 1978 — Another week speeds by and I feel we've not done enough. It goes slow this week because of problems beyond my control. Also, Maggie is under the weather right now, so I've done no night work this week.

Enough apologies. we have started the ceiling at the lowest points where it's the easiest to reach: the bathroom, Selena's room, back and front halls. Here's our system: salvaged 1x4s serve as supporting furring strips that run perpendicular to the old pine pole rafters. The ceiling material I finally settled on is 4x8 foot sheets of Armstrong Temlock at $5 a sheet. We're using big-headed roofing nails to tack the sheets up. What really keeps the ceiling in place is two-inch wide molding strips we sawed out of old barn wood. The final ceiling looks vaguely Tudor — but it reminds me more of the World War II era temporary housing for Naval pre-flight students and later, guys on the G.I. Bill at the University of North Carolina at Chapel Hill — a place called Victory Village where the style was '40s Barracks Utilitarian.

Sunday, May 20, 1978 — The last two weeks have been spent on the ceiling. (Picture that, if you can.) I did the living room single-handedly since Bill has summarily taken off for Nashville presumably to do something with his music. I brought all my old scaffolding in from the chimney work and constructed three units to reach the 18-foot-tall ceiling. But the ceiling material is so lightweight that the work was surprisingly easy. And the final product doesn't looks as rinkydink as I once feared it would.

The kitchen, as expected, has turned out to be a whopper of a challenge. The kitchen cabinets arrived yesterday as promised — and they are really lovely. Solid pine and stained with Minwax's Early American. But as we moved them in place, we hit a snag. My plumber (who has left for Costa Rica) didn't leave enough rise in the plumbing above the floor, so Joe Ryan will have to come and rig something special before cabinet-makers Ray and Dan can come back and bolt the blamed thing down.

Also, I've just learned that nobody wants to mess with custom countertops, so it's up to ol' Slick. As of today I've got 23 Things left to do.

Saturday, May 27, 1978 — I returned to the cabin in the evening after

The loft with banister and closet all but complete. And I've just finished the ceiling.

friend Jeanne Petillo got married on the front porch of her log cabin up Shingle Hollow way. I worked until 11:30 p.m. on high obtuse angles of the ceiling in bathroom and front hall. It was so quiet down here and I was concentrating so hard that when one of Jeanette's cats came mewing into the cabin she spooked me pretty badly. Earlier I had parked the car way up at the top of the hill — and with no moon at all, the night was dark as a coal pit. I walked out with the kerosene lantern swinging boogeyman shadows into the woods.

Sunday, May 28, 1978 — Joe installed final lighting receptacles and I acted as apprentice. We worked into the night to finish. The tile people come tomorrow to do blue and white floor covering in bathroom and kitchen. Also, we've ordered a chocolate brown Formica countertop which Joe has agreed to install.

Friday, June 2, 1978 — I finished the final touches of stucco work in the front hall and kitchen. The front hall is just flat-out spectacular.

I got a late start today because I had to wait until Jeanette and Maggie had finished putting down the fifth and last coat of polyurethane on the living room floor. It glows with a deep reddish patina that only old heart pine possesses.

A bit of excitement when 21-month-old Jon decided that the newly-varnished floor would make a great skating pond. He sneaked in the living room, fell in the wet polyurethane, and then made little footprints all over the porch and screamed bloody murder about getting cleaned off.

This evening after Maggie left, I spent hours putting up the front door. Four big butt hinges had to be chiseled in place, and the door itself is so massive that I can just barely grunt it around. It must weigh 200 pounds. But I hefted it into place, and even then had to spend ages sanding and chiseling both door and frame to get it to fit perfectly. But then there she was — one instant ancient door. The Old World look and weighty feel of it remind me of the German church doors on which Martin Luther nailed his theses.

Thursday, June 8, 1978 — My birthday. Nine-year-old Selena presented me with a card she'd written: "To Dad   Age 33   But Still Going Strog."

Going strog. That about sums it up.

Another cosmic birthday present: One day last week after a sweaty work day, I was taking a shower down at Bill and Jeanette's when the phone rang. No one was at home so I answered it. Turned out to be a thousand-to-one shot — it was Charlie Ratcliffe, to whom I had advanced a hundred bucks four years ago. "How's the cabin coming?" he asked. He was calling to see if I still needed him so he could come work off the debt.

The timing couldn't be more perfect as we seem to be entering the final month of finishing the cabin. So today he came and worked on the very difficult angle closet doors that had to be made out of old thick oak. Later we set out to build the front steps out of old logs left over from the door cut-outs, but instead in a burst of creative insight turned it into a semi-circular ramp instead. Later, I worked until midnight stuccoing the back hallway.

Sunday, June 11, 1978 — Four years to the day since we started this wildness. It is with the utmost satisfaction that I mark this anniversary. Graduation is at hand. Next week maybe?

Charlie and I had a zoomy day building the base for the giant deck. Two years ago I had started this with the base for a 14-foot, six-inch long deck, and today we doubled that size. It is 28 feet long and 16 feet deep, and supported by massive log pilings culled from door, window and fireplace cutouts from the original main cabin.

Charlie says that viewing the cabin from the driveway side "you'd never guess there was a durned dancehall on the other side."

The Charlie Ratcliffe Memorial Dancehall Deck changes the whole cabin in a way I never could have anticipated — extending both living rooms and kitchen out, into, and over the woods. Now, standing sunburnt and fuzzy-headed out on the corner of the deck's prow six feet off the ground, I can turn around and look back at the cabin — and lo, it's like seeing an old friend for the first time in four years.

There he is — this is the same eye-level, three-quarters perspective from which I viewed the old barn when it sat up on the road, squatting in the honeysuckle waiting for me. This is my dream's-eye view. Seen from here, the cabin no longer looms up awkwardly like an ungainly mama chicken trying to get airborne. Now, the cabin appears more like something imagined by the Disney Studio from the '40s — a place for Snow White to hole up; Heidi-snugglish yet not cutesy. My Swiss blood must have had a stronger hand in this design than I had previously conceded. The place has an authentic alpine look, yet not Subdivision Chalet.

Meanwhile, everything else is snowballing along. The amazing Joe Ryan has been so important to me these last few weeks that words fail to impart the full extent of the man's role. After finishing the final plumbing in the bathroom and kitchen, he has put down the kitchen countertop, a demanding job involving a 10-foot-long, horseshoe-shaped galley with drop-in stove at the end.

Yesterday, while I helped him glue down the countertop, I was alarmed to see him pick up my red gym "Yes Shorts" to wipe off the excess glue. I grabbed them from his hands like I was snatching my toddler out of a freeway.

"Thought it was a rag," Joe said flatly, nodding at the bedraggled, hole-shot faded red shorts.

"Well, it's not!" I retorted in mock offense.

"Could have fooled me," Joe replied dryly in his typically New England tone, turning back to his work.

My Yes Shorts. My beautiful red four-year-old Damn the Torpedoes, Full Steam Ahead gym shorts with the go-ahead motto sewn on the thigh. They were this white man's breach cloth since that first crazy summer of '74.

They've been patched, retreaded and reborn — a garment of historic proportions which I shall have bronzed or framed and hung in a place of honor over the fireplace where in my advancing age I can gaze fondly at them recalling four insane and completely soul-satisfying years.

A rag indeed. That, sir, is my diploma.

Monday, June 12, 1978 — I just heard on the news that builders in North Carolina say home construction costs this last year have gone up nine to twenty percent — and that in the Raleigh area the average cost for a 1,500-square-foot new home is $57,500.

We bought 50 acres for $15,000 and that's already paid for. I'll bet the house won't cost over $10,000, and we'll have the building material loan paid off in three years. Need I say more?

Friday, June 16, 1978 — Everything is in italics now, happening at 78 rpm. I feel like something out of a Mack Sennett blotchy black and white film, running around too fast and jerky for real life, but getting things done licketysplit.

Working at the newspaper this week, I look at my hands in wonder. How foreign and demure the typewriter keys seem by comparison to the things my methiolated hands have been manhandling this final week: logs and hammers and pickaxes to the point that my muscled right hand actually feels larger than my left. The keys feel like a child's toy and each one of my fingers feels as large as a fist. I knit my brow at the new hands I don't recognize; has their owner metamorphosed as much as they have?

Saturday, June 17, 1978 — Big time. Joe has finished and the kitchen looks fantastic. He only charged $283 for all his work over the past month, too.

Charlie Ratcliffe finished his week with me, and more than made up for that debt. When he was leaving and we were shaking hands, I thanked him for being so loyal — and then asked him how much I owed him for "tuition." Charlie stopped a second, and then there was that lovely grin that breached his easy-going face. We hugged. "This is one great house," he said, slapping me on the back.

Then later, while I was rushing about doing finishing touches, James Harmon's grading boys showed up to redo my terrible roads. So I hot-footed it up to the top of the hill and we walked the course for the switchbacks and talked about ditching and draining — and I liked the outfit so much I said let's go with it.

While I was showing the bulldozer operator, J.V. Biggerstaff, around the place, he allowed that he grew up nearby, but didn't say where exactly. I was going on about how we moved the cabin down from up on the road, and how we'd salvaged an old farmhouse up there too — when suddenly J.V.'s face lit up. "You mean the old house up the road?" He could scarcely believe it. "Why, I used to live there," he announced.

"Thought that thing would have rotted down years ago."

Turns out the maestro bulldozer artist lived in the Giving House when he was about 5 years old and that he vaguely remembers the old log barn he used to play in — the same "barn" to which he was now cutting a road. The irony of it: his upstairs bedroom had contributed some of the most perfectly intact wood for the new house. I took this as a good omen.

They did great work. And now after four years of grappling with road problems, our system is now graveled and suitable for normal vehicular travel.

Also today — with the final plumbing done, I fit the rubber pipe from the spring to the house intake valve and — well, there it was: we had water, as if by magic in all the faucets and shower and commode. The sound of the water rushing through the copper pipe for the first time felt like being inside a living being with the blood coursing through its veins.

Sunday, June 18, 1978 — A preliminary D-Day. We got a solid start on moving, starting with the living room of the Blue Shed — everything toted out in the newspaper van. This is thrilling, but we've got another week's paper to get out, and more packing to go before we hit Normandy Beach full scale. Both Maggie and I admit to being antsy. But with graduation at hand, I reckon we're entitled to a little "senioritis."

June 21, 1978: it's done, and ready for us to move in.

18

In Which We Live the Dream

"Behold, we count them happy which endure."
— *James, 5:11*

"If you done it, it ain't braggin'"
— *Dizzy Dean*

Wednesday, June 21, 1978 — I admit it. I've spent the last four years as a closet pioneer, a schizo editor/log cabinman. I've burned the candle at all three ends. I've OD'd on roofing, been stoned by a chimney, been floored, chinked, fiberglassed, block-busted, Hogwild every minute of the way.

People keep asking, "Would you do it again now if you knew how much time it was going to take?" I don't know if I can answer that accurately just yet. I can confidently say, "I wouldn't take nothin' for my journey now," (in the words of the old gospel song). But don't ask me that question today — because the reply would be a bumptious Yes!

For today was moving day. And today we began living the dream.

The last couple of nights we lay awake at night in the Blue Shed amid piles and boxes, sleeplessly and speechlessly planning the moving strategy, eager to be up at 4 a.m., yet knowing we'd pay later if we did. We fell into fitful frenetic sleep, shot through with zany dreams.

There's Mrs. Olsen, and I'm chasing her around a giant Folger's Coffee can the size of a log cabin with a cedar shingle roof.

Now, Charlie Ratcliffe and I are on that roof, and we're hammering down shingles one by one, only this time they're hamburger buns instead of wood shingles.

Then there along the crest of the eastern ridge of Laurel Mountain, we notice the ominous silhouette of hundreds of Nazi soldiers bearing down on us.

But instead of attacking, the Nazis break into a chorus line, a vaudeville kick line, jackboots in step-knee, step-kick and they're singing some silly tune: "It's so much fun to be a Nazi/ It's downright hotsy-totsy/ With Hitler, Rommel and Goerring/ Oh, we love to go a'warring."

And by now the Broadway musical Nazis are swinging from the trees onto my scaffolding, and...

"You're snoring," Maggie punched me awake. "I'm sorry but you were snoring."

"Maggie, you're not going to believe what I just dreamed," I tried to tell her. "The cabin was a coffee can and the roof was hamburger buns and we were being attacked by singing Nazis and..."

"Babe," she hushed me. "I believe you've got cabin fever."

And why not? What better reason to be delirious? The cabin is finished. Today we moved in. Two sentences I've waited four years to say. And even now it sounds like a grand fib.

My imagination can be forgiven; I'm in overdrive — physically and mentally supercharged by reality's finality. The calendar for June looks like Ike's pre-D-Day engagement schedule. Marshaling forces, coordinating men and materials, getting the right people to do the right thing at the right time at the right place. Four years of building rises now to a smashing crescendo of hammer beats and circular saws whining in a crashing Wagnerian final chord.

We started out incorrectly loading the U-Haul truck with dinky stuff in the back. Then, around mid-day, friend Robert Miller, the Incredible Black Hulk himself, shows up — and it turns out Robert's granddaddy was a professional mover, so Roberto knew a thing or two about the art. Plus, Robert has muscles, so that just the two of us were able to load the piano, as well as other big items into the truck. Friend Harold Gleaves (who will be renting the Blue Shed) came by to lend a hand too, so by midafternoon we were in great shape.

After loading breakables into Robert's VW van, we threw dog, cat, and kids into the Volvo and went tearing out to Hogwild forever. I took the big U-Haul down the new road triumphantly; the whole trip in was like a victory parade.

We unloaded stuff all afternoon, with Robert and Harold being great help. The only real problem we encountered was the big refrigerator which was so large it required that I knock one of the Zonk supports in the kitchen temporarily out of place, and then I had to also tear out some sheetrock to make room for the double-door critter.

By 5 p.m. it was all off the truck: beds in rooms, Jon's crib in the back hall, the piano in the living room, the washer and dryer in their cubbyhole in the bathroom, the big china hutch in the living room, a million boxes of books and stuff stacked about and Maggie trying to make some sense of it all.

After our moving friends left, we worked on into the evening — and other friends dropped by to wish us well — potters Claude and Elaine Graves, painter Bill McCullough and wife, Nancy; and Bill Byers with a housewarming bucket of fruit and wine.

It is St. Han's Day, the longest day of the year.

The Longest Day indeed.

We were so busy and happy we couldn't think of joining the annual party down at the Big House until we got more done at our little house.

A most imperial full moon sailed up over Hogwild — mystical and welcoming us tonight. And for once a homey sight.

Around midnight after the kids were asleep, we walked hand in hand down to Bill and Jeanette's — two grubby lovers strolling past party-goers strewn about the Big House's lawn — many of whom didn't recognize us at first — then when they did, were probably wondering what those crazy Lauterers where doing up at this hour of the night. We walked together as if in slow motion.

And then a homebound moonlight stroll to savor. How many nights have I seen the cabin in the moonlight when the thought of living there was always a distant dream, never the reality. Now the dream is real. We are living the dream.

This is Commencement.

We showered and went to our bed in the lovely loft bathed by sheer blue moonlight angling in through the windows. We slept as if dead.

Friday, June 30, 1978 — The last day of June. The most exciting June of my life.

This week we have been Settling In. Stowing away things in long-awaited places. In a fit of unfettered creativity, Maggie has hand-painted a line of bluebirds and green vines ringing the kitchen wall above the windows, and it echoes the design in the tiles done by the Graves. She was worried that I wouldn't like it, but I think it pulls the kitchen together graphically. I've built some shelf units upstairs out of salvaged pine, and now I'm doing bookshelves downstairs. It's all so easy now that we live here.

Tonight I am seated at the kitchen table, the old black Royal which has accompanied me down to the old picnic table for so many days of writing these past four years is now jouncing on the old round claw-legged table.

The first words a man writes in the house he's built should, one supposes, be grandific; Spenserian sonnets dripping adjectival verbosity, splendid verse worthy of literary journals. Not an American Legion baseball story. Clunk goes reality.

Sunday, July 2, 1978 — It is raining just now, a good hard afternoon trashmover. I love rain here; I always have. But it is so different from before — when the rain seemed more of an adversary, threatening me with isolation, hemming me into the Holler, forbidding me to leave.

But now here is home, and I needn't leave. It may seem like a subtle difference but it's a profound shift. It's taken a simple thing like a summer shower to make me feel deep down the enormity of the change: for four years we were camping at the Blue Shed and then in one blitzing day we've turned our world on its ear.

Saturday, July 15, 1978 — Over the last ten days I have truly savored working out here, even more than I did the last four years because I've not had to leave it each night. I'm about finished with the giant deck. I went down to Shelby and got my salt-treated pressurized lumber direct from a supplier — 380 square feet of 2x6s for only $245. It would have been almost double that elsewhere.

I have built my railing and supports out of logs and chestnut poles lying about. One monstrous 4x6 oak beam that I'd been saving for four years turns out to be exactly the right length — 14' 6". This is amazing to me since this beam was found just lying in the old barn, not a part of the structure as such. I kept waiting for a place to use such a hunk of wood; that's what I mean about the wisdom of waiting for the fullness of time to reveal itself to you. This piece of wood didn't want to be cut for lesser purposes, and I had the good fortune to wait until the very end of the building process to be illuminated as to its real purpose. The deck is now finished and it all but doubles our living space. What a joy it is.

Thursday, Aug. 17, 1978 — Tomorrow night we are having our sure-enough housewarming. We've gone into hyperdrive getting everything "party-ready." This has served as a great excuse to finish all those dibbydabby things you tend to let slide once you live in a place: I laid a rock front walkway in an S-curve, did the finishing touches on four units of bookshelves, eight feet tall and about 16 feet long. Then I mowed the "yard" such as it is, and also mowed the woods all the way down to the creek and around Selena's Playhouse. In addition, Bill has Bush-Hogged the Soccer Field, and it makes the whole place look like a wilderness estate.

Friday, Aug. 18, 1978 — If, upon my death, the Infinite asks me if I would like to re-live one single day, I believe it would be today.

From start to finish, it was moving, total, consuming bliss.

The Housewarming was one big Stroke. And for me personally, it like a final chapter in the incredible journey begun a distant four years ago.

The place looked like it was ready for a visit from Better Homes and Cabins. About 20 folks gathered out on the deck in the cool of the idyllic August evening with fiddling and singing, my mortar pan converted to an icy tub full of beer, plates of Maggie's lasagna, Jeanette's biscuits and a full August moon blooming over Laurel Mountain to add to the kerosene lanterns. A sweet soul-satisfying evening.

People who'd watched us build this place over the years just walked around gawking... asking questions with unbridled admiration: "How'd you do this?" "How'd you do that?" If the building process hadn't taken so long I think I would have been embarrassed.

So many of the people had had a hand in the cabin one way or the other: Rick (our solar energy zealot) and Maureen Pratt from Tryon; Clayton Spencer and fabric artist wife, Joyce, who are building their own place over that way too; artist Bill McCullough, who helped with Selena's Playhouse, said, "I wouldn't have missed it;" *This Week* reporter Joy Hoagland who's heard her fill of my trials and tribulations; fiddling carpenter Joe Ryan; my co-editor, Ron Paris, (who's been through a lot of sacrifice on account of this place) and wife, Janice — as well as Hogwilders Stuart and Cindy Byers (who are building like crazy now too); Burwell Byers; and of course Bill and Jeanette — our blood brother and sister of the land. I could be mistaken, but ol' Bill and Jeanette looked as happy and as profoundly satisfied as did cabinfolk Jock and Maggie, who, after everyone left, stayed up very late just talking and savoring.

Ready for the housewarming. On the deckside, Jon's got his liquid refreshment ready.

19

Life and Death Within the Dream

"In my end is my beginning."
— Mary Queen of Scots

Monday, Sept. 21, 1978 — My old dog died today. My old Ruby the Weather Dog, Ruby the big red alligator. The best dog in the world. I found her dead this morning in her bed in the VW bus. We heard her moaning last night, but it was Sunday night, and I didn't suppose it was serious. Must have died in her sleep last night. From what I still don't know. Could I have saved her? I guess I'll never know.

She was simply the best dog in the world — a big, smelly red-haired hunk of a mutt, a glad-hearted mix of golden retriever and St. Bernard. The size of a small fire truck and the same color. As gentle and kind as the day is long. Great cinnamon sad eyes set in a handsome head charred with a square black Bernard's muzzle that in its 11th year had assumed a distinguished gray frosting.

She was our first baby. We got her in Chapel Hill for a song, and Maggie christened the black-faced pup after a lovable and wise old black cook she'd known. Ruby was always at her happiest when riding in the car, hanging out of the window just as far as gravity and inertia would allow, and — on several occasions, inertia won out -- and the poor dog was sent skidding and tumbling into ditches.

But usually she had the wind and momentum under control. All through my log cabin misadventures Ol' Rube accompanied me on salvage runs, on supply trips — no matter how tightly packed the VW van, there was always room for the dog: wind-swept, red and black hair rippling like a Kansas wheat field before the storm, her great proud head jutting into the wind, mouth open and her tongue lolling, she appeared to enjoy eating the rushing air.

In her old age she had grown neurotic about storms, hence her nickname, the Weather Dog. She could always sense a storm brewing, and when one did blow in, she sought out our company, staring wildly at us with those reddish eyes and panting: is it okay? Will it go away?

When the children came, she was never jealous like an older sibling might be. Instead, she assumed her new duties as Mother Protector and canine nanny.

She loved to swim, hence her other moniker, the Big Red Alligator. She had the peculiar talent of drinking water from a spigot. She never bit anyone. She stayed close to home. She barked at the right time and with proper decorum. She was a true lady.

Digging a grave for a 75-pound dog is an act of contrition. For all the times I forgot to feed her, left her out in the cold, treated her mean, didn't return the devotion she gave us, or didn't take her moans more seriously.

The Soccer Field's rocky soil yielded reluctantly to my shovel. I worked up a towering sweat and was glad for the absolution, even though this morning was cool and the sun had yet to burn away the dew on the lace doily cobwebs in the grass.

We cried aloud for Ruby. Laid her there in the meadow. Cried for our first baby to die. I told the Lord that Ruby was due a rabies shot. Hot tears blinded my eyes as I shoveled the dirt onto the red fur below. "I can't look!" I wailed, blubbering like a child.

Maggie, holding little Jon, turned away across the meadow to pick flowers for the grave. Even little Jon, who called her "Boo," sensed the solemnity of the passing. Waving to the purple asters on the grave, he said, "Bye Boo."

I can't shake the feeling that this is the end of an era.

October, 1978 — I am building again. Rebuilding. Recovering and healing. A replacement dog has been acquired to try and fill Ruby's paws. But "Lucy" is a yappy nervous foxhound and I openly resent

The new Woodshed and a young woodbutcher at work.

Southern Living comes to our house and turns the kitchen into a studio.

her intrusion. But we do it for the kids, who have their own needs. Selena adores Lucy, but every time I look at her cross-eyed the little hound shrinks in terror and pees on the floor.

For my own rehabilitation, I have salvaged and moved the little Woodshed from town, rebuilding it here for a nice tall, two-story barn where we can store stuff, keep our wood dry and give me a place to chop and saw in the dry of the little barn's overhang. The whole thing was built pretty much from Giving House and cabin left-overs.

During the last two weeks we've been visited by miraculous visitors from the outside world. First, a team from *Southern Living* called Maggie wanting to do something on mountain cooking in an appropriate setting. So photographer John O'Hagan and assistant foods editor Linda Welch came out to the cabin and staged this elaborate country breakfast under all sorts of studio lights. It won't be published until next fall, but we still felt like stars.

Then not long after that, I got a call at the paper from an Art Boericke saying he wanted to come check out the cabin for a new book he was doing on owner-built homes.

"You mean *the* Art Boericke?" I exclaimed. He was delighted and slightly embarrassed that I actually knew who he was. I rushed home, and told the old Californian that his original book had been one of the main influences pushing me to build the very house he had now come to see.

206 *Hogwild*

The biggest log house to go up on Hogwild yet — Burwell has himself a log mansion.

Brother Bruce's log house goes up to the north of me. That's Ben hanging on the back of "Will Hogwild."

What wonderful irony. And there was photographer Barry Shapiro whose pictures had freed me up to build Hogwildly. While Art talked with Maggie, I trailed around with Barry talking shop as he shot the house.

We discovered we both had the photo-documentary urge. I told him I had a new photo book in the works, something on old folks from the Southern Appalachians. Barry said he is working on a photo book of his own: "San Francisco Transvestites I Have Known," he said. And we laughed at the outrageous juxtaposition.

Hogwild grows.

This month Bruce Byers has finished his cabin up to the north of us. He built a "kit" cabin with 6x8 logs, so he did it pretty much singlehanded, except for help from his girlfriend, Toni Shell, a college student at UNC-Asheville. Bruce is an ideal neighbor to occupy the headwaters of Silver Creek. We've backpacked together, worked together on conservation projects, and I respect him tremendously for his convictions, gentle temperament and perseverance.

His younger brother Burwell had a monstrous cabin-raising recently which took every able-bodied soul — and then some — from Hogwild and beyond. This new log structure, right beside the little one we all helped raise three years ago, will be a log mansion when he gets it done.

There, I just caught myself; "when he gets it done," I said.

How pleasant to use that phrase about somebody else's place.

The Charlie Ratcliffe Memorial Dancehall Deck gets put to good use by Selena and friends doing disco.

Sunday, Nov. 1, 1978 — Sourwood Holler is alive with comings and goings. Overhead the sky leaps up to take precedence as the foliage slips away. Again, we can see clear down to the creek, to the footbridge and beyond to the little girl's fanciful Playhouse built years before while the cabin was still more barn than house.

The leaves, for which the leaf-lookers journey hours from cities to glimpse, put on a brief private exhibition. Even after several tentative warning frosts, my friends the paper wasps zoom drunkenly about in summer's last hurrah, looking for refuge. I know that feeling. This year I've found mine.

Home at last. "We Raise the Song of Harvest Home."

In the cerulean blue globe that cleanses itself each autumn I watch a patient red-tailed hawk describing figure eights, and there's a furiously-flapping barn owl caroming through the gathering dusk.

And a final treat: glimpses of that massive migration in progress, the confident, fragile monarch butterflies slipstreaming down from the north. One by one the aeronauts flutter purposefully high over the cabin, guided by some unseen benevolent hand ever due-south toward their Mexican winter hideaway.

Monday, Nov. 2, 1978 — Having just read some ancient Chinese philosophy, I feel moved to put my contented frame of mind into simple, declarative observations. Perhaps if Lao Tsu had been a Hogwilder he'd have put it something like this:

I am seated in my wicker chair in the darkened study of the new little barn I have just finished. Outside dusk has fallen on a fine Indian Summer's day. Above me the stars wink on. Below me the woodpile is dry and tall. Out of the salvaged windows I see the familiar outline of the house I built. The rock chimney's granite jaw rises solidly against the warm beckoning windows of home. Inside the woman is cooking supper in a country kitchen. The daughter is peacefully in her room doing her homework. The two-year-old is banging on a cooking pot with an egg-beater.

Ah, is this not happiness?

20

Homecoming

> *"No Epilogue, I pray you, for your play needs no excuse."*
> — *Shakespeare*
> *"A Midsummer-Night's Dream"*

> *"Still Crazy After All These Years."*
> — *Paul Simon*

Thursday, May 14, 1992 — The passing of almost 20 years have been kind to "Old Tom," as the cabin has come to be known. Would that I had come through that period as unscathed. I am sitting on the silver-gray deck as a perfect spring dusk creeps down Sourwood Holler. Below, a swollen Silver Creek is chortling from two days of rain. Above, an almost full moon peeps over the silhouetted piney east ridge. The sweet smell of May is tinged with the acrid smell of wood ashes — ashes not from a wood stove but from a house fire.

Homecoming indeed. The very term implies that you have a home to come back to. Only the heroics of friends and the Green Hill Volunteer Fire Department makes this simple moment of serene creekside contemplation possible.

It's hard enough for me to grasp that this cabin is pushing 20, and that the bright golden cedar roof Charlie Ratcliffe, Bill Byers, and I lay down in 1974 should have turned to so much tinder just waiting for a dry spell, an errant spark, and a malicious wind to burst into a conflagration. In short, my roof had turned into a disaster waiting to happen.

On Friday night, Jan. 31 of this year, all those factors combined. And it wasn't until he heard the roar of the flames that my house-sitting friend, Tony Napoli, realized what was happening.

Phoneless, he sent a visiting friend sprinting up the hill to get Bill, who a couple of years ago moved into Gene Ham's log cabin. Bill called 911 and then set out for the cabin to help.

Meanwhile, Tony was so adrenalated that he practically vaulted onto the porch roof and then up onto the main roof where the fire was raging through the dry shingles beside the kitchen chimney.

He'd dragged the garden hose up there with him, but when he turned it on, nothing came out. The old thing had crimped and tangled below, cutting off the water. But Tony wasn't about to let that stop him, and so he ripped off his sweater and began beating on the flames.

Tony said later that as he was fighting the fire he had a "little chat" with the Almighty, not making any promises about lifestyle changes, but acknowledging that some divine providence would be mightily appreciated right about now.

And right about then, he heard the sirens. By a stroke of good fortune (how about more divine providence) the Green Hill boys only knew where the cabin was because of a grass fire the week before down at the nearby Big House.

About this time, Bill arrived at the top of the road and saw the flames in the Holler. "Good-ness, Jock," he exclaimed later, "I wasn't worried about your cabin — I figured that was a goner — I was worried about the woods."

Bill said the fire was so hot that zephyrs shot off the burning shingles like a prairie fire in a stiff wind. So hot the Styrofoam insulation crumpled like squashed ants. So hot the aluminum roof flashing melted in misshapen silver blobs, and the asphalt roofing paper flowed and dripped like melted wax.

The firemen arrived, pausing briefly at the top of the road to wonder if they could get their big trucks down my little driveway. And then throwing caution to the winds, they barreled down into the Holler, threw out their hoses, leapt on the roof, and doused the flames.

The roof had burned through at the hottest point, flaming embers falling on Tony's bed below in the loft. Seeing this, the firemen rushed in the house, busting one of the loft windows so they could hurl the smoldering mattress out on the deck below where they could hose it down without fear of water damage to the house.

In fact, they put out the fire with so little water damage that to this day I remain astonished. Oh sure, there is soot unending and little things to be put right, but it's piddling by comparison to what could have happened.

This fire was a close thing — a dodged bullet, a narrow escape, disaster averted. The other day when I went up to the station to thank the firefighters, an assistant chief who had been on my roof told me frankly, "Another five minutes and you coulda' wrote that place off."

Another five minutes and almost 20 years of loving that place — not to mention the four years of building and dreaming — would have been nothing but ashes, rubble, and memories. To say I was lucky is shallow — more like gifted, blessed, reprieved.

In February, Tony cleaned up and patched the roof and found himself a new place to live so I could move back for a summer of writing and reclamation. And in March, my Brevard tennis partner and roofing contractor, Dennis Newman, who I call "Dr. Roof," gave the old cabin a brand-new fireproof fiberglass-asphalt shingle roof, the same pale golden color of the original cedar shakes.

So here I sit tonight in the gathering spring evening as the smells of sweet May assail me — beauty bush and flowering wild cherry — good woodsy loamy earth smells. The silver dollar in the sky reminds me how close I came to having to draw comfort from the old Buddhist expression:

My barn
having burned to the ground
I can now see the moon.

Friday, May 15, 1992 — I arrived at Hogwild yesterday from Penn State where I have just completed my first year as an assistant professor of journalism with the dream job of building the School of Communications' photojournalism program. From Hogwild to Penn State, from six families on 300 acres of woods to a Big 10 campus with 39,000 students: as Maggie and I used to say, "'It's a long way to Tipperary.'"

When I drove down the old driveway, I felt a jumble of conflicting emotions: scared of seeing the fire damage, confronting the reality and pain that I had left all this... and yet I couldn't deny my delight at being able to live here again — even if it's only a summer writing retreat.

OK, I said to myself, turning the Jeep Cherokee onto the well-graded and deeply graveled "Hogwild Blvd.," I know this sentimental journey is going to be bittersweet, a minefield strewn with historical relics sure to blow up in my face. I steeled myself for the inevitable ambush. Duly, the journalist noted the changes: Bill and Jeanette's bodacious mailbox of the '70s ("Jeanette and Wm. J. Byers are Hogwild") had been replaced years ago by a cute model put up by the Floridians who had bought the place. But that mailbox, made to look like either a tiny chalet or a monstrous birdhouse, was now rotting, falling apart, and a wren was nesting in its ruins.

Yet that was nothing compared to my mailbox, which was all but on its knees, the old support log totally gone to rot. Together, the two mailboxes looked downright decrepit.

Continuing down the road: no trace of where the Giving House stood, and a field of daisies occupied the original log barn site of 18 years ago this summer. Pausing at both places, again I felt the kinetic energy of the ground over which I'd worked and dreamed.

Just down from there, with a mixture of humor and sadness, I saw the wrought iron and brick gate the Floridians had built. Two imposing brick columns were topped by ornate lamps. I wondered, what is this — a country club? The entrance to The Addams Family estate? Talk about an anachronism. Turning toward my humble driveway, I grew philosophical. I knew I was going to get ambushed, and here it was.

But I was unprepared for the next one. My narrow driveway, taking off to the left just above the ostentatious gate, had become overgrown with little pines. Maggie with her beloved hedge trimmers would have had a field day here, I chuckled to myself at the thought of her happily "zapping" the loathsome underbrush.

And then there was Tom. If anything, he looked more at home in the Holler than I could have imagined. The growth of trees we had planted or worked to save now nurtured the cabin in return. The knee-high Colorado blue spruce planted the Christmas of '78 stood 16 feet tall, obscuring a quarter of the house from view.

I stood on the lip of the hill, looking at the scene, trying to take it all in. This is not a dream, not a memory, not a photograph. I am here and

Jon cooking in Tom's kitchen, summer 1992.

Selena writing in the living room, summer 1992.

this is now, As proof, I reminded myself, there's the new roof looking good.

Then my heart leapt. There stood Ruby. In the deep shade of the chimney loomed a red shaggy apparition. I stood perfectly still and consciously made myself start breathing again. Think, I commanded. It must be a neighbor's dog which by sheer coincidence is an over-sized golden retriever. Walk, I told my wooden legs. Beat, I told my stumbling heart.

Suddenly, the spirit-dog bounded happily out to me — just the way Ruby would have. I fell to my knees, threw both arms around this totally strange dog in complete trust, burying my wet face in its mane — and we rolled around the ground like children tussling. Thus did Ruby-reincarnate welcome the prodigal back to Hogwild.

When I had recovered sufficiently, I noted how through the last 12 years the renters had taken good care of Old Tom. Clearly they, too, had loved it and respected the workmanship.

After being away for so long and having lived in so many modern houses, my first impression of the cabin was one of authenticity, of texture and rich colors, earth-tones of rough gray logs and lighter gray chink, of native stone and silvered barn siding, of white stucco walls and deep russet gold of the old pine floors, doors, and cabinets.

How could I have left this place? It's a treasure, a family legacy beyond reckoning. Walking around in Tom, I realized the rightness of living here again — and especially of passing this magic kingdom on to the children. Indeed, not surprisingly, Selena, the little Fairy Princess of Slickey Mountain, had turned into a poet. After finishing college and working in Greensboro for a year as a para-legal, she announced her plans recently to move back to Hogwild this summer to work on her writing. What could be more perfect in the fullness of time? And "little" Jon, little no more, but now a strapping 6'2" rock 'n' roll drummer boy of 15, thinks this place is "cool" and "awesome" and wants to soundproof one of the rooms for practice sessions. I have to chuckle at how he is turning out. Born to drum, fetal "LizaJon" must have been going for the bass drum pedal all those times he kicked Maggie.

Sitting on the far corner of the deck (the prow, I used to call it, for the angle juts into the woods like a ship bravely plowing through the green breakers), I was again immersed in the aura of the place: quiet yet zooming with energy and peace for me. This is one of my main creative "hot spots," a mystical power base that recharges and regenerates me.

If this is so, then what has kept me away for so long? Why has it taken the Bedouin journalist this long to take this sentimental journey home?

After years of emotional questing, I think I have the answer now. Because when we are hurt and pull away, we tend to pull too hard. At least I did. It reminds me of sophomore high school biology when we studied the reproduction of spirogyra. The way the algae made more of itself was by a process called "conjugation," during which the single cell would stretch and elongate as if its two disparate poles were fighting to get out of the same grocery bag. The cell looked like it was caught in a taffy-pull until the two parts lurched violently apart in a microcosmic separation. Similarly, when our marriage ended, I had pulled myself away from Hogwild and the mountains. Pulled too hard, perhaps. But such were the needs of healing.

"Rooo-by," I crooned softly to the not-Ruby dog snoozing in the sunlight at my feet. "Hey, big fella, how 'bout a walk?" I stirred out of my lethargy as the dog jumped to his feet at the magic word, "walk."

First, I inspected the homestead, cataloging the deferred maintenance. Selena's Playhouse was still standing, though the vines had taken over. The Woodshed needed a new roof in the front. And the Smokehouse, bless its heart, was still leaning at its gravity-defying angle. As I was checking out the Smokehouse's crumbling block foundation, I spied something green and glinting buried there in the leaves. A glassy shine. It could only be one thing. And so it was — a pony-sized Rolling Rock bottle. I could date it exactly to January 1976. What a joke on me. I'd stuffed the foundation with my empties, thinking how archeologists from 2420 would find my "time capsule" from the '70s — and now look who turns out to be the explorer in his own past!

Setting off on my pilgrimage around Hogwild, I paused at the top of the road where I had a clear view of Old Tom below in the Holler; it was right here that the cabin had received its anthropomorphic handle. During a 1982 visit when I had Jon, 6 at the time, we were driving out of the place — and at this exact spot, the little boy had asked pointedly, "Dad, who named Ruby?"

"Maggie," I replied.

"Who named Lucy?" he asked of our current dog.

"Selena did," I responded.

"Who named Sourwood Holler?" he wanted to know.

"I reckon I did," I said, wondering where all this was leading.

"Yeh! And you named Obie and the Blue Shed, and Selena got to name Slickey Mountain!" Jon blurted as if this was something to which he'd given considerable thought.

" — And I haven't got to name nothin' nothin'!" He sounded definitely left out of the process.

"Well," I said, taken aback by the injustice of it all, "what would you like to name?"

"What's the cabin's name? You got a name for the cabin?" I was conscious of the boy looking at me carefully, his blue flashlight eyes boring in.

"No. Reckon it's just always been The Cabin." I was really intrigued now. "What do you want to call it?"

"TOM," Jon announced seriously with scarcely a pause. "The cabin's name is Tom."

Did I laugh? I don't remember. I only know I was instantly delighted at the notion of "Tom" — perhaps with the affectionate prefix "Old" added for cadence and respect. Old Tom the mountain man who brought us the rocks from high off the Blue Ridge years ago. Old Tom Bombadil from the Hobbit's tale. Tom Tom the Piper's son/Stole a pig and away he run. An agelessly wise and roguish boy of an old man. Something of an irascible curmudgeon in wood. A cabin named Tom. It was perfect. How could the little boy know how well he had chosen?

So, on up the drive I went, wondering at all the changes that had happened to me, to Maggie, to Bill and Jeanette, and to the folks of Hogwild. Who all was still here? What did their houses look like? How many kids did they have? What had happened to the Hogwild dream? I had only inklings of some of the answers.

Whereas in the mid-'70s our group was bonded by common goals, by the end of that decade, some of us were swept away from Hogwild by other equally compelling currents. The Hogwild dream became not so all-consuming, and some of our lives seemed at the mercy of forces we couldn't readily understand.

Three families left and broke up, and three families stayed together and on the land.

The three that left, ironically, were the original three Hogwild families: Byers, Walther and Lauterer. The three families who stayed on the land and stayed together were the three younger Byers brothers who had joined us later.

Of the original three, the Walthers had the least to do with Hogwild. They did some reforestation, but that was all. Later, they divorced, and Rosemary married the town mayor and moved away. Jim moved south and went into newspaper management.

In 1982 Bill and Jeanette sold their amazing house to the Floridian couple, who quickly added a front yard swimming pool — the very height of gauche, it seemed to us Hogwilders at the time.

Standing at the top of my road, gazing through the gate down at the Byers' former Big House, I recalled how Bill and Jeanette had always been driven by great dreams. When they left Hogwild, it was for a new dream: to head west, buy land or an old house and start a ski lodge. However, the trip west proved that land was either inaccessible or exorbitantly priced. Returning to the South, they bought and renovated an old house in the ski village of Banner Elk and converted it into a bed and breakfast ski lodge. Meanwhile, Bill worked on his MBA at Appalachian State University.

Although by 1983 Maggie and I had already gone our separate ways, we were stunned when we heard of their split. Why is it that your best friends' breakup is more shocking than your own?

Jeanette stayed to run the inn, and Bill moved to Charlotte to sell commercial real estate — philosophically about as far a cry from Hogwild as one could possibly get, we thought.

Well, I mused, turning to hike up the main road toward the old log barn and Giving House sites, it was no more revolutionary than what happened to the Lauterers. We grew in different directions. First, I began veering away when our weekly newspaper, *This Week*, went daily in late '78. It wasn't long until I was captivated by the siren song of starting another weekly.

We left Hogwild in March 1980, having lived here less than two years — after working on it for over four years.

The new weekly in neighboring McDowell County was only a marginal success, and after two very difficult years, the recession of 1982 delivered a coup de grace, and we sold out and broke up. As irony would have it, I returned to the cabin alone and lived there for a year of healing, before going to teach at the University of North Carolina at Chapel Hill, Brevard College, and then Penn State. During this time, 1982-92, in order to make ends meet, we rented the cabin to a series of house-sitting friends.

Maggie was recruited by the Asheville *Citizen-Times* where she became a respected feature writer, photographer and columnist with the touch of an Appalachian Will Rogers. She subsequently joined WLOS-TV in Asheville, continuing her unique personal style of journalism, winning many awards for her work. A couple of years ago, she married a wonderful man, Zack Allen, a former reporter turned director of the Western North Carolina Nature Center, and together they bought and restored a delightful old *Waltons* style place in East Asheville. For their wedding present, I gave them six weekends of bust-a-gut labor at the old farmhouse.

Throughout all this, we agreed the children were our primary concern — that they feel loved through the separation — and so we kept that family feeling in spite of our distance and personal needs. We stayed close. We nurtured them through the split the best we could and through it all remained good friends.

So as I stopped by the field where we'd found the vine-covered log barn back in the winter of '74, I realized books and houses have a way of never ending where you think they will. Maybe things don't exactly end at all. They go right on living in spite of our best attempts to put down periods. Is that what Tom was trying to tell me all along? That humanity tends to see time as a horizontal line with significant dates and events marked off with short perpendicular strokes: birth, school, marriage, children, career, retirement, death — all neatly projected, plotted on your own chronological ruler. If Old Tom taught me anything, it's that change and growth always and forever are spiraling upward and outward. Maybe from now on I'll see life not like a yardstick, but rather a chambered nautilus.

As if to underscore that thought, the big red dog and I found ourselves at the lip of the road leading down to Gene Ham's log cabin, "Kyoto Knob," inhabited now by none other than the former King of Hogwild, the architect of my dream, Bill Byers, a.k.a. Will Hogwild, himself. Now there's a neat, tight circle. My buddy on the land had returned at last to his music, his original great love. A fine singer-songwriter and guitarist, Bill had done several tapes in Nashville recently, and I think they're quite good. Moreover, he's doing exactly what he wants to do. Plus, I hear he's bought a piece of Hogwild adjoining me atop Laurel Mountain. Calls it "Toehold." What could be finer than for Will Hogwild to be a'building again?

Of the Byers' children, I learned that son Ben, 17, lives there with Bill; Hannah, Selena's faithful childhood playmate, is a rising senior at UNC-Asheville majoring in history and psychology. The youngest, Will, a rising eighth grader, is with Jeanette, who remarried and moved to the famous resort community of Southern Pines and then to Beaufort, S.C. Busy as ever, she is studying computer graphics and selling real estate.

And what of the other Hogwilders? I hear Gene Ham is a public school teacher in south-central Tennessee where he and his brother bought one of his old family homeplaces. Maybe he'll come visit us this summer. Every time I think of Gene, I am reminded of our fraternal society, the Knights of the Okra, whose motto is "Semper Gumbo."

Bill McCullough, the superb artist who lived across the river and who could also have made his living as a rockmason, moved back to the Low Country after inheriting the family farm near Georgetown, S.C. I learned he's become a successful portrait and landscape painter. Living in Charleston, he also teaches art at the Gibbes Museum.

Charlie Ratcliffe, my right-hand man, who I'd seen a couple of times in Chapel Hill while he was visiting friends, was a successful and happy interior decorator specializing in designer wallpapering. Now I hear

220 *Hogwild*

During a summer 1992 reunion, clockwise from front, Will, Ben, Gene Ham, Bill and Hannah Byers at Gene's cabin.

he's based in nearby Waynesville as a builder. He's remarried and has a young family.

Mike Thompson, my mentor as both carpenter and teacher, wisely returned to his first love — teaching. After working for years as coordi-

Homecoming 221

nator of the gifted and talented students program at McDowell High School, he had two books on the subject published. In 1990 he moved to Jackson, Tenn., where he married a third time, and now works with the academically gifted at the University School of Jackson, still inspiring young minds to greatness, I'll warrant.

Tom Cowan, the wonderful watercolor artist who, for a time lived humbly in Bill and Jeanette's horse barn and later in a school bus parked out here, has seen his fortunes change for the better, too. An inheritance has allowed him to paint, travel, and teach as he pleases. Most recently he is setting up his own studio in Gatlinburg, Tenn.

Art Boericke and Barry Shapiro, who researched and photographed Tom after we'd finished, published their work in a handsome book, *Handmade Houses: the Natural Way to Build Houses*, 1981, Delacorte Press.

The multi-talented carpenter Joe Ryan, who helped me in so many ways in the final completion of Tom, still lives near here up on Clark Road, where he has a busy workshop.

And Keith Harrill? I'd learned that "Too-Tall" (who helped me build the Smokehouse that crazy night in '76) had become a builder in the county. Well, that fits.

By the time I'd mulled over all that, I'd arrived at Bruce and Toni's place — but I hardly recognized the "cabin," for it was a cabin in its core only. Its size having quadrupled, the place was more like a homestead,

Bruce and Toni Byers with sons, left to right, Zachary, Elliott, Alexander and Samuel.

Stuart and Cindy Byers and children, clockwise from front, Judson, Caleb, Etta and Rachel.

an estate where slick had overtaken rustic, with a classy white concrete driveway looping attractively up the hill and into a spacious turn-around play area bordered by a log garage housing a Volvo station wagon, a playhouse, basketball court and late-model VW touring bus. Off to the side they'd built a swimming pool and, farther up the hill, a lighted regulation tennis court.

The reason for all this became quickly apparent. Bruce and Toni had four boys: Zachary, 11; Elliott, 9; Samuel, 5; and Alexander, 3 in August. At present they were all out shrieking with laughter as they bounced on the trampoline below the house. Inside, the house had grown handsomely with stucco walls, exposed beams, stout casement windows and the designer colors and wallpapers that Jeanette and Maggie pioneered. Bruce and Toni dressed as if they shopped from Lands' End or L.L. Bean, and they wore it well, striking me as content and stable.

Toni remembered how in 1977 when she was still in college and helping Bruce build the cabin on weekends, they had a falling-out. Toni said, "I remember thinking, 'Shoot! he's going to live here — and after all the work I've done,'" She said, grinning across the room lovingly at Bruce.

Burwell and Jeannie Byers and sons, left to right, John, Thomas and Henry.

As the dog and I left and loped up the hill toward the next brother's place, I encountered a new log cabin on top of the ridge facing west. The brothers' parents, Jim and Louise Byers, had had a charming weekend get-away log cabin built on old "Father Austin's View."

At the homes of Stuart and Burwell, the growth had been equally exponential. By contrast, in the mid-'70s those two free-spirited guys had started by camping out at Hogwild in a teepee. Now look at their places. The original cabins of 15 years ago served now as cores for places that cried out for photo spreads in *BH&G* or *Southern Living*. Yet the ambitious additions melded harmoniously with the original structures.

Stuart and Cindy had four children: Judson, 12; Etta, 10; Rachel, 7; and Caleb, 4. Their multi-leveled house seemed to go on forever. Outside, the house was complemented by at least four log outbuildings, each one a masterpiece, grouped around a paved drive with a pool tucked discreetly behind one barn.

As the day lengthened into dusk I clumped down the trail to Burwell and Jeannie's. There I found that they had three boys: John, 10; Thomas, 7; and Henry, 5. Burwell had turned his place into something of a log mansion; there's just no other word for it. Huge expanses of thermopaned glass mounted in soaring gothic gables, cathedral ceilings, barnboard, rockwork and exposed beams — not to mention several more log barns. It occurred to me that the brothers had collected log barns the way some people do fine art. Hogwild had become a living museum of Southern Appalachian log structures. And the three remaining families had taken the design feel of early Hogwild and enlarged grandly on the theme. Unless I miss my guess, I counted no less than 15 log cabins and barns during my circuit of Hogwild.

They all still worked for the family propane gas company and had made a good thing of it. Burwell and Stuart had bought the eastern bottomlands and far hill when logging had gotten too near, thus expanding Hogwild by another 100 acres. As I left the last brother's home, (minus the dog, who'd turned off in the woods somewhere) and traipsed down to the river road, I knew I was under the spell of a new but enduring Hogwild. The magical homecoming day was stretching into a sweet spring evening, and Stuart's easy, broad grin kept looming in front of me like a Cheshire cat. "Yeah," he had said with satisfaction, arms crossed over his chest, "we're getting dug-in."

I hit my land down where the river road runs just above the rushing Rocky Broad, and there I was greeted by a musky bouquet of river scent percolating the spring dusk. As strongly as that fecund aroma, the thought came over me: Hogwild lives! There was a continuity after all. The '70s dream had changed, just as the cabins had grown in size and style, but the commitment to the land and to each other was very much intact in the '90s. The three brothers and their wives had taken the baton passed by Hogwild's founders and had carried very well in our absence.

Now Bill Byers is back and building. I'm living in Tom, who survived his trial by fire. And Selena is coming to live here. Even the Floridians, who, after a decade, have become "Hogwild-ized," are a part of this place, too. They don't seem so different from us now as they once did. Have they become more like us, or have we become more like them?

I couldn't help wondering at the contrast between the '70s and now. Had we all turned into the YURPies I'd joked about back then — Young, Upwardly-mobile Rural Professionals — with central heat, pools, minivans, Jeep Cherokees, VCRs, camcorders, and retrievers? Even the irony of my spending the summer writing on a state-of-the-art Apple Macintosh in a log cabin was not lost on me.

As I hiked on along the overgrown trail back toward Old Tom, I realized how much of a community this place has become. Just look at the number of kids here now. Why, Hogwild could easily field its own Little League team.

At the brow of Sourwood Holler I was greeted by the sound of a homecoming committee in full voice. Down in the Soccer Field a whole vibrant flotilla of bog frogs, spring peepers, were chorusing their wheezy welcome.

Then, as if on cue from some master stage-lighting technician, over Slickey Mountain there billowed a May full moon, casting its golden luminescence upon the cabin standing sturdily in the Holler below. I stood transfixed, as if hearing my name called faintly but unmistakably. It was Sourwood Holler calling me home. I was caressed by the knowledge, as sure as the moonlight, that Hogwild would always beckon, and that I would always answer, "Yes!" The ultimate affirmation of what I had dreamed of, worked on, lived for, lived through — as well as left behind, returned to, reclaimed, and reconciled, could be reduced to three simple letters: Yes!

Another moonrise over Sourwood Holler as Hogwild beckons me home.

A Hogwild reunion 18 years after: St. Han's Day, summer 1992.

Credits

Grateful acknowledgment is made to the following for permission to reprint previously published material:

Mudslide Slim by James Taylor (c) 1971, SBK Blackwood Music Inc. and Country Road Music Inc. All Rights Controlled and Administered by SBK Blackwood Music Inc. All Rights Reserved. International Copyright Secured. Used by Permission.

Woodstock Handmade Houses (c) by David Ballentine and Robert G. Haney, 1974, Woostock, N.Y. Used by Permission.

Excerpts of "Ski Bum Turns Earth Mother," article by Lew Powell, *The Charlotte Observer*, 1975, Used by Permission.

Still Crazy After All These Years (c) 1974, Paul Simon; and *Slip Slidin' Away* (c) 1977, Paul Simon, Used by Permission.

Popeye (c) Reprinted with Special Permission of King Features Syndicate, Inc.; N. American English Language Rights.

Spiderman (c) Reprinted with Special Permission of Marvel Entertainment Group, Inc.; N. American English Language Rights.

Colophon

Hogwild's body text is Palatino 10 point, the title face is Goudy Old Style. The book was designed on the picnic table in Tom's kitchen using a Macintosh IIsi with Pagemaker. Photos were taken with Nikon F-2s with Kodak Tri-X film.

The author, mid-way through construction of Old Tom, summer 1976.